Also by Roger Ellman:

* THE ORIGIN AND ITS MEANING

ON THE ORIGIN OF THE UNIVERSE AND ITS MECHANICS,
THE MECHANISM AND ORIGINOF INTELLIGENCE,
AND THE IMPLICATIONS FOR THE INDIVIDUAL AND SOCIETY

* ON THE NATURE OF MATTER

THE ORIGIN OF THE UNIVERSE CREATED MATTER FUNDAMENTALLY WAVE IN
NATURE, NOT PARTICULATE

* THE PHILOSOPHIC PRINCIPLES OF RATIONAL BEING

ANALYSIS AND UNDERSTANDING OF
REALITY, TRUTH, GOODNESS, JUSTICE, VIRTUE, BEAUTY, HAPPINESS, LOVE,
HUMAN NATURE, SOCIETY, GOVERNMENT, EDUCATION, DETERMINISM,
FREE WILL, AND DEATH

* GRAVITICS

THE PHYSICS OF THE BEHAVIOR AND CONTROL OF GRAVITATION

* THE TROUBLE WITH THE HUBBLE LAW

AN ALTERNATIVE TREATMENT OF REDSHIFTS IS EVIDENCED BY
FOUR INDEPENDENT OBSERVED COSMOLOGICAL EFFECTS.

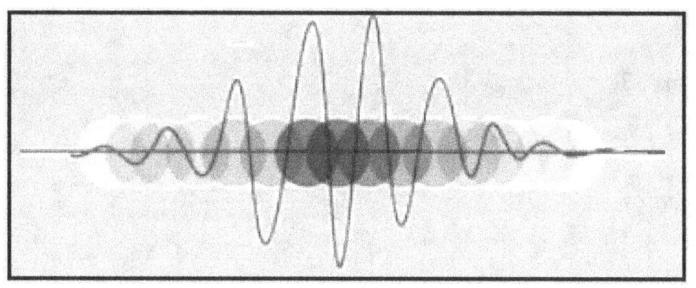

RESOLUTION OF THE "SPOOKY" PROBLEMS OF QUANTUM MECHANICS

THE WAVE - PARTICLE DILEMMA,

WAS ESSENTIALLY DECIDED IN FAVOR OF THE PARTICLE CONCEPT BY THE COMBINATION OF PLANCK'S TREATMENT OF BLACK BODY RADIATION AND EINSTEIN'S TREATMENT OF THE PHOTO–ELECTRIC EFFECT.

IT IS HERE SHOWN THAT THAT CHOICE WAS A MISTAKE. THE WAVE NATURE OF PARTICLES AND LIGHT RESOLVES EINSTEIN'S "SPOOKY" ENTANGLEMENT PROBLEM AND THE "WEIRDNESS" OF QUANTUM MECHANICS.

ROGER ELLMAN

Cataloging Data

Ellman, Roger (1932-   )

Resolution of the "Spooky" Problems
  of
Quantum Mechanics

Resolution of the "Spooky" Problems
of
Quantum Mechanics

Library of Congress Control Number: 2019900825

Published by: The-Origin Foundation, Inc.,
                 1401 Fountaingrove Pkwy.
                 Santa Rosa, CA 95403, USA

                 707-537-0257

                 http://www.The-Origin.org

ISBN 9781795070294

CONTENTS

i

ABOUT THE AUTHOR

The-Origin Foundation, Inc. is a non-profit organization founded to foster independent scientific, mathematical, and philosophical research.

The author of *Resolution of the "Spooky" Problems of Quantum Mechanics"*, Roger Ellman, is the General Director of the foundation.

Roger Ellman has published over fifty professional papers on topics ranging from physics, cosmology, and astrophysics to artificial intelligence and mathematics.

He has presented some of his papers to conferences of / at:

The American Physical Society [APS], .
The American Society for the Advancement of Science,
Cambridge University, United Kingdom
The Library of Alexandria, Egypt
The Russian Academy of Natural Sciences, St Petersburg
The Hungarian Academy of Sciences, Budapest
A Science Conference in Shang Hai, China

He is author of five books in addition to the present one: *Resolution of the "Spooky" Problems of Quantum Mechanics"*.

His education includes graduate studies at Stanford University after graduating from West Point, the United States Military Academy.

In classical physics the principles of "realism" and "locality" are fundamental, in effect axioms.

Realism is the principle that all objects must objectively have a value of any of their measureable characteristics before any measurement is made and independent of any measurement made.

Locality states that an object is only directly influenced by its immediate surroundings. For an action at one location to have an influence at another non-contiguous location, something in the space between the locations must mediate the spatial separation.

Quantum Mechanics rejects those principles, especially in its phenomenon of "entanglement".

Quantum entanglement is a quantum mechanical phenomenon in which the states of two or more objects have to be described with reference to each other, even though the individual objects may be spatially separated. This leads to correlations between observable physical properties of the systems.

As a result, measurements performed on one system seem to be instantaneously influencing other systems entangled with it in spite of their spatial separation.

The "spooky", as Einstein called it, aspect of this is the violation of locality, the action at a distance with no intervening mediation of that separation. Einstein and others sought to show that there were other factors operating that did mediate the separation, but they were not successful. The problem remained, "How did one particle or system of particles communicate with another that is spatially separate ?"

It is here shown that such a communication cannot exist and is impossible. Therefore the interpretation of the effect is mistaken or if not the effect would be ***supernatural magic***. The situation is analyzed and the error demonstrated.

The logic of the presentation follows on the second next page.

LOGIC OF THE PRESENTATION

On the Resolution of the Problem of Realism

1. There is no cause nor mechanism for the following contended QM effects:
   - Not for superposition of states,
   - Nor for its particles existing only as a probability wave function,
   - Nor for particles' change from probability to specific existence
     being initiated solely by an act of observation in some form, and
   - Nor for particles knowledge about conditions physically beyond their ken.
2. Analysis of the Double Slit Experiment shows that:
   - All of the quantum behavior observed is accounted for by classical physics;
   - Therefore, there is no need for the unsupported QM explanation.

On the Resolution of the Problem of Locality and Entanglement

1. For the QM effects there <u>must</u> be communication between the parties such that:
   - Every particle must continuously broadcast its quantum state,
   - Each entangled particle must have an identification to that effect, and
   - For each entangled pair there must be mechanism enforcing correlation.
2. Each particle does universally broadcast communication of its state but that lacks:
   - Identification of its being entangled, and
   - Correlation enforcement mechanism.
3. That communication lacking and a second with that first being impossible, either:
   - The entanglement and its effects do not exist, and are a misunderstanding, or
   - Entanglement is valid, an actual example of, by definition, ***supernatural magic***.
   - That latter conclusion cannot be.

On Quantized Angular Momentum and the Stern-Gerlach Experiment

1. As the temperature increases in the quite hot silver vapor of the furnace the collimated beam of silver atoms becomes partially ionized, positive ions.

2. The electrons lost by the ionized silver atoms flowing in the collimated beam are temporarily acquired by other silver atoms which become negative ions.

3. Each rapidly moving ion in the beam is an individual electric current and per Ampere's Law generates a magnetic field and consequent magnetic dipole on each ion.

4. In brief the effect of as if the silver atoms have "spin" and quantized angular momentum is actually due to their loosely bound outermost electrons and their migration within the silver atom beam resulting in individual positive and negative silver ions' local Ampere's Law current and consequent magnetic dipoles. The quantization is singly ionized [one electron] positive and negative silver atoms.

   - Thus here quantized angular momentum is not an inherent "natural property" of particles rather its apparently appearing is really a classical electric effect.

REFERENCES

[1] R. Ellman, *The Origin and Its Meaning*, The-Origin Foundation, Inc., http://www.The-Origin.org, 1997..

[2] R. Ellman, *On the Nature of matter*, The-Origin Foundation, Inc., http://www.The-Origin.org, 2017.

[These may be downloaded in .pdf files from http://www.The-Origin.org.htm]

PART I -- THE PROBLEM
The "Spookiness" of Quantum Mechanics

The problem is that quantum mechanics apparently exhibits behavior and calls for accepting beliefs that contradict the accepted facts and behavior, the axioms, of the pre-quantum mechanics world of Newtonian classicism. To some, including Einstein, this is "spooky".

- Realism and Locality.

- The Nature of Quantum Entanglement

- The "Spookiness" of Quantum Entanglement

Realism, Locality and the "Spookiness" of Entanglement

REALISM AND LOCALITY

In classical physics the principles of "realism" and "locality" are fundamental, in effect axioms. They are as below.

The analysis begins with those two axioms because Quantum Mechanics directly and overtly denies the validity of both. Resolving that problem, that contradiction, is the objective of this work. The two axioms are comprehensively valid. The problem is to determine how the problem came about and demonstrate how it is resolved.

Realism

Realism is the principle that all objects must objectively have a pre-existing value of any of their measureable characteristics independent of any measurement that is made and before the measurement is made. The measurement cannot and does not create or initiate the value.

Locality

Locality states that an object is only directly influenced by its immediate surroundings. For an action at one location to have an influence at another non-contiguous location, something in the space between the locations must mediate the spatial separation.

Both axioms seem perfectly rational to us. To us their proof is in their statement. Of course, things are as they are without any human intervention or consent. We are not gods.

And we rely completely on the concept of cause and effect. It would be weird magic for something to be acted upon from a distant spatial separation with no accounting for that intervening space.

QUANTUM ENTANGLEMENT

Quantum entanglement is a quantum mechanical phenomenon in which the states of two or more objects have to be described with reference to each other even though the individual objects may be spatially separated. This leads to correlations between observable physical properties of the systems.

3

For example, it is possible to prepare two particles in a single quantum state such that when one is observed to be spin-up, the other one will always be observed to be spin-down and vice versa, this despite the fact that it is impossible to predict, according to quantum mechanics, which set of measurements will be observed.

As a result, measurements performed on one system seem to be instantaneously influencing other systems entangled with it. [But quantum entanglement does not enable the transmission of classical information faster than the speed of light.]

By definition, entanglement is a type of correlation among two or more particles (or other systems). One finds that they are entangled by measuring them and finding that the results are correlated. However, there are many subtleties. In measuring these systems, one is apt to destroy the very entanglement sought. Also, it cannot be relied on that the correlations will be strong enough to differentiate them from classical correlations. So, in practice, one knows that particles are entangled because you prepared them in a proven way. Often you can look for so-called entanglement witnesses, which are large-scale consequences of entanglement.

The more precise statement of quantum entanglement is as follows.

In quantum mechanics, if two particles are in a state such that there is a matching correlation between two "canonically conjugate dynamical quantities", quantities like position and momentum, whose values [by Schrödinger's definition] suffice to specify all the properties of a classical system, they are termed as being "entangled".

Experiments have been conducted the results of which have been interpreted as instantaneous communication of a such 'canonically conjugate' dynamical quantity from one particle to the other, the communication exhibited as a responsive change in one particle due to an introduced change in the other particle.

That is a case of measurements performed on one system seeming to be instantaneously influencing other systems entangled with it.

Einstein famously said that he refused to believe in quantum entanglement's "spooky action at a distance". The "spooky", as Einstein called it, aspect of this is the violation of locality, the action at a distance with no intervening mediation of that separation. The "action" has been validly observed and proven so that Einstein and others sought to show that there were undetected other factors, "hidden variables", operating that did mediate the separation. They were not successful. The problem remained, "How in those cases did one particle or system of particles communicate with another spatially distant ?"

PART II -- THE WRONG CHOICE
The Dominance of the Particle Theory of Matter and Light

The wave - particle dilemma is that light exhibits behavior and properties that lead to characterizing it both as a wave and as a particle. This dichotomy was essentially resolved in the late 19th Century - early 20th Century in favor of the particle form of light. Also, the issue of a wave nature of matter particles having arisen with the discovery of matter waves, the dichotomy in the form of matter was likewise resolved in favor of particles.

That all seemed essentially reasonable. It is much more natural to us to conceive of matter particles, with all of their forms and activities, as like little round balls than as some form of wave.

But, the choice in favor of particles and neglecting waves was unfortunate in that it is the wave nature of matter that leads to the solution of the problem of Einstein's "spooky" action at a distance.

This Part II presents the evolution of this dilemma and of the dominance of the particle view of light and matter, the consequent early foundations of quantum mechanics, and the resolution of the dichotomy in favor of waves showing, also, that it is impossible for light to be particles.

- The Wave-Particle Dilemma

. The Matter Wave Problem

- Planck's and Einstein's Quanta

- Analysis of the Photon

The Wrong Choice:
The Matter Wave Problem and the Photon

THE WAVE-PARTICLE DILEMMA

The wave - particle dilemma is how to resolve that light exhibits behavior only explainable as it being an electro-magnetic (E-M) wave phenomenon yet light also exhibits behavior that would appear to be only explainable as it being of particle nature. The evidence for the wave nature of light is extensive including the wave behaviors of: reflection, interference, refraction, diffraction, frequency, wavelength, polarization, and so forth and the highly successful Maxwell's Equations describing it.

Then phenomena appeared that seem to require a particle nature of light hence its particle name, "Photon". Those phenomena were the failure at short wavelengths of the theoretical Rayleigh-Jeans law of black body radiation, then the photoelectric effect and the line spectra of gases. This evidential wave-particle duality led to the concept of the photon as a particle in the form of a "wave packet". But, the particle nature of the photon still has a number of problems.

A radially outward wave in free space tends to spread out as it propagates, but its deemed particle photon wave packets must be considered as staying together like a particle. The E-M wave front is continuous, but a front of propagating particles involves the particles' moving radially from the source with the distance between particles increasing with distance from the source and nothing in the spaces between.

E-M radiation is produced by acceleration of charge and must produce E-M propagation that is spatially symmetrical to the charge's motion, but the particle theory requires that the radiation travel away from the accelerated charge as a specific particle in some specific direction without symmetry. The next particle may be in another direction, the next in a third, and so on, so that a large number of radiated particles exhibit a dispersion pattern somewhat like that which the wave field would, but that still is behavior that is inconsistent with the wave aspect.

When waves encounter an impenetrable barrier with an aperture in it, the portion of the waves that encounter the aperture and pass through propagate from its far side as if it were a new source of radiation, that is in all available outward directions. Particles in such a circumstance, that is those particles which encounter the aperture instead of the barrier, should simply continue traveling in straight paths. If the particles (because said to be packets of waves) were to propagate in the fashion of waves from the aperture they would either have changed from particle to wave, or have each subdivided into numerous particles, or have cooperated by leaving the aperture in random directions simulating the behavior of a wave field

THE MATTER WAVE PROBLEM

In the early 20th Century (1924) DeBroglie proposed that, since light, which was then considered to be a purely wave phenomenon, had been found to appear sometimes to exhibit particle behavior; perhaps matter, which was accepted as being particle in nature might sometimes exhibit wave behavior. DeBroglie reasoned that, the wavelength of a photon being equal to Planck's constant, h, divided by the photon's momentum, the same relationship should apply to a particle of matter -- it having a wavelength of h divided by the particle momentum.

The reasoning was as follows. First considering a photon, its energy is

$$(2\text{-}1) \qquad W_{wave} = h \cdot f$$

where h is the Planck Constant and f is the frequency.

The energy equivalent of a mass, m, is

$$(2\text{-}2) \qquad W_{mass} = m \cdot c^2$$

where m is the mass and c is the speed of light.

While the photon's rest mass is zero it has kinetic mass corresponding to its energy. If the photon equivalent mass, m, actually appears as a wave its energy as a wave must be the same as its energy as a mass. Therefore

$$(2\text{-}3) \qquad W_{mass} = W_{wave}$$

$$m \cdot c^2 = h \cdot f$$

$$m = \frac{h \cdot f}{c^2}$$

$$= \frac{h}{\lambda \cdot c}$$

and, finally,

$$(2\text{-}4) \qquad \lambda = \frac{h}{m \cdot c} \qquad\qquad \text{[solving } (2\text{-}3) \text{ for } \lambda]$$

$$= \frac{h}{\text{photon momentum}}$$

by recognizing that momentum is defined as the product of mass and its velocity and the velocity of the photon is c.

DeBroglie hypothesized that the wave aspect of a particle of matter should have an analogous wavelength, λ_{mw}, that should be

$$(2\text{-}5) \qquad \lambda_{mw} = \frac{h}{\text{particle momentum}} = \frac{h}{m \cdot v}$$

This suggestion of DeBroglie was soon verified by Davison and Germer who obtained electron diffraction patterns and found that the observed wavelengths of the electron matter waves corresponded well with DeBroglie's formulation.

10

At that point one would think that the duality of matter, as of light, was established and that extensive further investigation of matter waves would have resulted. But that was not the case and the reason was a fundamental problem that could not be overcome – the matter wave frequency.

If one reasons, analogously to the derivation of λ_{mw}, that the kinetic energy of the particle of matter should correspond to its matter wave frequency, f_{mw}, as

$(2\text{-}6)$
$$f_{mw} = \frac{W_k}{h}$$
$$= \frac{\frac{1}{2} \cdot m \cdot v^2}{h}$$

then the velocity of the matter wave is

$(2\text{-}7)$
$$v_{mw} = \lambda_{mw} \cdot f_{mw}$$
$$= \left[\frac{h}{m \cdot v}\right] \cdot \left[\frac{\frac{1}{2} \cdot m \cdot v^2}{h}\right] = \frac{1}{2} \cdot v$$

a result that states that the matter wave moves at one half the speed of the particle. That is obviously absurd as they must move together each being merely an alternative aspect of the same real entity. For them not to move together would be as absurd as for the particle aspect of light to move at a different speed than its wave aspect, the photon not arriving coincident with the $E-M$ wave.

It is no help in resolving this difficulty if relativistic mass is used (as it should be in any case) since the same mass appears in both numerator and denominator of equation $(2\text{-}7)$ where they simply cancel out.

It is also no help to hypothesize that it is the total energy, not just the kinetic energy, that yields the matter wave. Such an attempt attributes a matter wave to a particle at rest. It also gives the resulting matter wave velocity as c^2/v which has the matter wave racing ahead of its particle. No, the two must keep pace with each other since they are the same thing merely looked at in one or the other of two alternative ways.

It was the inability to resolve this problem that led to the loss of interest in matter waves and essentially the end of further inquiry with regard to the wave aspect of matter. That resulted in modern physics in the complete dominance of the particle form of light and matter, a development initiated by the work of Planck and Einstein.

PLANCK'S AND EINSTEIN'S QUANTA

The amount of energy radiated at various wavelengths from a material body varies with the temperature of the body (e.g. a piece of metal as heated hotter and hotter changes in the apparent color of its glow from dull red through orange to bright white). Experimental observations of this consistently show a characteristic pattern as in Figure 2-1, below: low energy magnitude at small wavelength, followed by a peak around a wavelength that depends on the temperature of the radiating body, further followed by a tapering off as wavelength further increases. If a theoretical curve is calculated based on E-M wave theory, the Rayleigh-Jeans law of black body radiation, and compared to the measured actual results a discrepancy appears. The theoretical curve increases without limit as wavelength becomes smaller instead of peaking and then declining toward zero.

This conflict of theoretical and actual behavior was resolved by Planck. He found that if the theoretical curve was derived upon the supposition that the radiated energy was given off in minute bursts of waves, bursts later called "quanta", rather than as continuous waves, and that each minute burst has the energy of the product of the radiation frequency and a constant, then the theoretical curve matched the actual curve for the given temperature. That relationship is equation *(2-8)*, where f is the frequency of the radiation and h is the (then new) universal constant, subsequently named *Planck's Constant* the value of which has been found to be about $6.626\ 070\ 040(81)\ x\ 10^{-34}\ J\ s$.

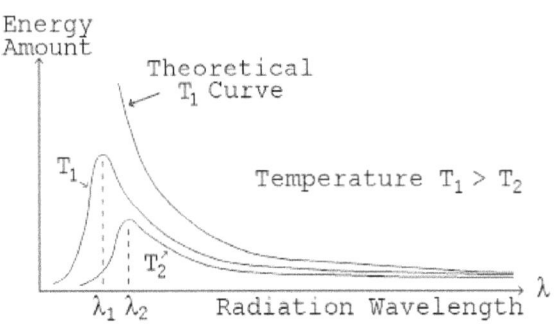

Figure 2-1

(2-8) $W = h \cdot f$

Einstein and the Photo-Electric Effect

Under suitable circumstances, it is found that when light or other E-M radiation of sufficiently high frequency shines on or encounters a material substance then electrons are given off by the substance. This *photoelectric effect* is the operating principle of television cameras, xerographic copiers, etc. The normal expectation would be that one would have to wait a shorter or longer time, depending on the intensity of the light, while it delivers enough energy to free the electrons, a heating up period so to speak. On that basis any E-M radiation should produce some electrons from the substance that it encounters if given enough time.

But, that mode of behavior is not the case. Experimental observations show that there is no heating up time, no apparent energy accumulation. Electrons are liberated by the incident light immediately if they are to be liberated at all. There is a threshold frequency, however, below which the light never releases electrons, at and above which electrons are always released, and then at which the rate at which electrons are given off depends on the light intensity.

The threshold frequency is different for different substances on which the light is shined. Furthermore, the electrons are given off with various individual energies, but the maximum energy of the released electrons depends directly on the light frequency. Figure 2-2, below, depicts this photoelectric effect behavior for different substances. The slope of all such lines turns out to be the same, as depicted in Figure 2-2 below. Furthermore, the slope turns out to have the same value as Planck's constant, h, the constant that Planck found necessary to explain black body radiation.

Einstein explained this behavior by postulating, similarly to Planck's assumption for black body radiation, that the light travels in packets of energy each containing the energy of equation *(2-8)*, $W = h \cdot f$. These packets of light energy were given the name *photons* and they were assumed, like matter particles, to be a discrete package traveling in one specific direction.

Einstein's hypothesis was that if a photon that is part of the incident light and that encounters an electron in the substance has enough energy W due to its frequency f so that the photon energy is greater than the energy binding the electron into the substance, then the electron will be released. Photons at frequencies below that threshold would not have enough energy to free an encountered electron. A photon of energy greater than the threshold would not only release the electron but would impart its excess energy to the electron as kinetic energy of motion. The rate at which electrons are released would depend on the rate of photons with time, which corresponds to the intensity of the light.

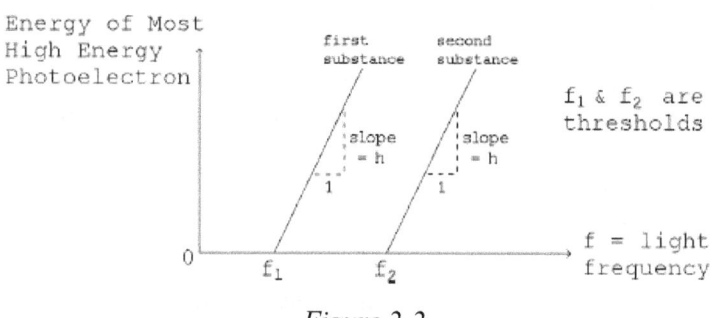

Figure 2-2

The Effect on the Foundations of Quantum Mechanics

The beginnings of Quantum Mechanics was distorted by the mistaken understanding of Planck's solution to the problem of a misfit of the classical Rayleigh-Jeans Law describing black body radiation. The mistaken understanding was the assumption that the quantization which successfully corrected the Rayleigh-Jeans curve represented the energy of actual particles. That was an over-assumption from the data. The data only showed that the energy transfer in black body radiation, a transfer from the hot radiating body mass to the propagating radiation, occurred in discrete amounts, quantized packets of energy change / transfer in amount $W = h \cdot f$. It makes no statement about the form of that radiation being like a particle.

Likewise, Einstein in explaining the photoelectric effect deemed that the incident light had to be in the form of particles. For Einstein: yes, discrete amounts of energy, $W = h \cdot f$, as with Planck, but also those in "packets" of some kind traveling like discrete matter particles in one specific direction. However it is impossible for natural light to be a discrete mono-directional particle; it must be a wave with a smoothly continuous lateral wave front, as follows.

ANALYSIS OF THE PHOTON FROM ITS GENERATING SOURCE

To resolve the problem of the nature of the photon it is necessary to go first to the constraints on what a photon is as they are imposed by its source, the transition of an atomic orbital electron from an outer to an inner orbit which transition must fit and match to, the following requirements.

- The transition is a change from the initial state to the final state as in part of a single cycle of an oscillation. It is not a changing to a different state and then returning back to the original state as in a full cycle of an oscillation.

- To avoid an infinite rate of change, which is impossible, the transition must be a smooth variation, without any sudden "jump".

13

- The resulting radiation exhibits all of the characteristics of Maxwellian electro-magnetic wave and is at one simple frequency, the photon frequency. It therefore must be in the form of a simple sinusoid.

- The theory of information in communications shows that at least a sample every half-cycle of an oscillation is required to specify it sufficiently. Therefore, at least a half cycle of the photon oscillation is required to specify it.

Therefore, the photon must be in the form of a half-cycle of a sinusoidal function of time. Beyond that, the following.

- In general, magnetic field is directly proportional to the velocity of the moving electric charge producing that field. Therefore the magnetic field of the photon is directly proportional to the transitioning electron's velocity. Since the photon magnetic field must be a half-cycle sinusoid the transitioning electron's velocity variation must be of a half-cycle sinusoid form.

- The electron velocity must vary in accordance with the above from the stable velocity of the initial orbit through a period of increase and ending in the stable velocity of the final orbit. (The potential energy lost in the move to a lower orbit appears half in the emitted E-M radiation, the photon, and half in the increase in electron kinetic energy due to its greater velocity in the inner orbit).

The combination of these factors results in the specification that the photon must be a half cycle of a pure sinusoidal type variation behaving as in the following figure 2-3.

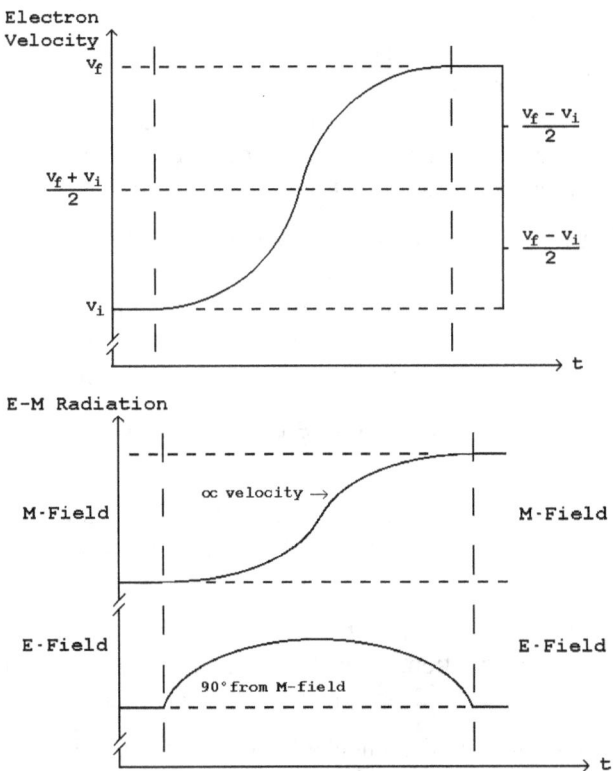

Figure 2-3 – The Orbit Transitioning Photon Generating Electron Behavior

14

The electron, traveling from its initial outer orbit to its final inner orbit with its velocity gradually increasing in a sinusoidal manner as in Figure 2-3, follows a path as illustrated in Figure 2-4 below, and emits an E-M wave field in "doughnut form" as in Figure 2-5, below, relative to its instant-by-instant varying vector velocity direction at each instant of the transition.

The peculiar shape of that field because of the directional orientation of the "doughnut" swinging through a substantial portion of a full circle according to the path of the electron's orbital descent causes the propagated E-M wave to contain the requisite form, angular momentum and energy for causing an encountered orbital electron elsewhere to be elevated to a higher orbit equivalent to the higher orbit that the electron previously descended from.

The propagated E-M wave burst contains and transmits both that energy and that angular momentum.

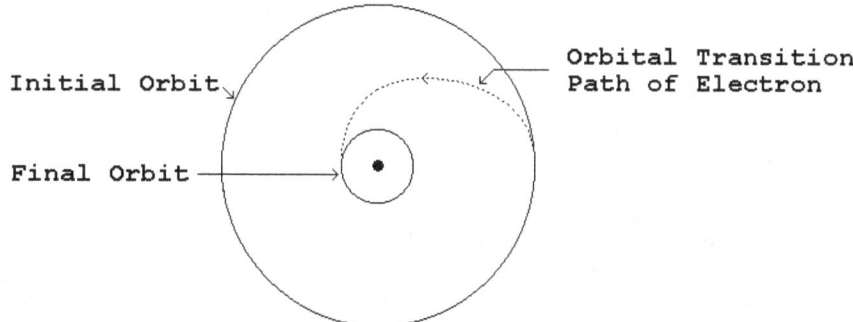

Figure 2-4 – Typical Electron Outer-to-Inner Orbit Change Path [Schematic]

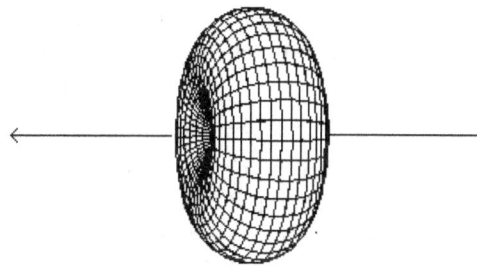

Figure 2-5 – Instantaneous Electro-Magnetic Radiation Pattern [doughnut shape] of a Single Electron Traveling as Shown.

Knowing the time duration of the electron's orbital transition relative to the electron orbital period is helpful in visualizing the process. The orbital transition takes place in the time of one-half cycle of the photon's frequency. Letting "D[xx]" symbolize "the duration of xx" then

$$(2\text{-}9) \quad \text{D [orbital transition]} = \text{D [photon]} = \frac{1}{2 \cdot \text{photon frequency}}$$

Let n be the orbit number, an integer representing the number of matter wavelengths in the orbit. The electron orbital velocity is proportional to $1/n$. Its matter wave frequency is proportional to $1/n^2$. The photon frequency is equal to one-half the difference between the initial and final orbits matter wave frequencies. From those, the duration of the orbital transition in terms of the duration per orbit in the final orbit is

$$(2\text{-}10) \quad \frac{\text{D [orbital transition]}}{\text{D [final orbit]}} = \frac{n_{initial}^2}{n_{final}[n_{initial}^2 - n_{final}^2]}$$

For some selected orbit transitions the results are as in Table 2-1, below.

n_i	n_f	Transition/ final orbit	n_i	n_f	Transition/ final orbit
2	1	4/3	4	3	16/21
3	1	9/8	5	3	25/48
3	2	9/10	6	3	36/81
4	2	4/6	7	3	49/120
5	2	25/42			

Table 2-1 – Typical Electron Outer-to-Inner Orbit Transitions

The transition takes place in $1-^1/_3$ final orbit periods at most [$n_i = 2$, $n_f = 1$], and in less than *1* final orbit period for most cases [$n_f > 1$]. The orbit transition and photon emission take place with the electron traveling a significant portion of an orbit around the atomic nucleus in all cases.

Looked at another way, the transitioning electron travels in the range of on the order of *180°* to *360°* of a circular orbit. Its path being curved its direction at instants during the transition changes by on the order of *180°* to *360°*. Such a path is completely incompatible with a particulate photon being emitted in some one specific direction by that transition.

Since 1960, when the first laser was made and operated, there have been two different forms of light. "*Natural Light*" The primary source of light is the transition of an orbital electron of an atom from a "higher" "stable" orbit inward to a less high stable orbit as just above addressed. It is impossible for such light to be unitary mono-directional particles. It is a form of wave that is generally of broad wave fronts and tends to spread out in space. Photons of such light are half-cycle sinusoidal waves in bursts of total energy of $W = h \cdot f$.

"*Coherent Light*" An important but much less ubiquitous form of light is that generated by a laser. A laser is a device that emits light through optical amplification based on the stimulated emission of electromagnetic radiation. The stimulated emission initially produces the above primary source kind of light but the amplification process results in the light becoming coherent. Coherent light is a beam of photons, particle like light waves, again half-cycle sinusoidal waves in bursts of energy $W = h \cdot f$, that have the same frequency and waveform and are in phase. Only a beam of coherent laser light is able to largely resist spreading and diffusing.

It is clear from all of the above that for "natural light" it is impossible that it travel outward away in a single specific direction as required for the photon-particle hypothesis. That role is restricted to "coherent light". Furthermore, there are other problems with the "natural light" particle hypothesis.

Perhaps the greatest other problem with the "natural light" photon-particle theory is as follows . The wavelength of light is in the range of 10^{-7} meters . Atomic dimensions are on the order of 10^{-10} meters so that if a photon is to contain wavelength data relevant to the light that it represents, it must then have dimensions that are on the order of $10^3 = 1000$ times the size of an entire atom let alone than the size of a much smaller orbital electron. Clearly this is completely at variance with the photon-particle explanation of the photo effect and line spectra of atoms.

16

For example, in the instance of absorption of a photon-particle causing the raising of an orbital electron to a higher orbit or to free of its site there would be, relatively speaking, a *football size photon* interacting with a *sand grain size atom*, the football-photon managing to focus its action solely on one *germ size electron* in the *sand grain size atom* <u>without disturbing any of the rest of the atom</u>.

The photoelectric effect has the same problem. For a photon-particle to eject an electron from an encountered metal material means a *football size photon* interacting with a *germ size electron* in the material composed of *sand grain size atoms* without disturbing anything other than the particular *germ size electron*.

THE CORRECT PHOTOELECTRIC EFFECT

Einstein received the Nobel Prize for his explanation of the photoelectric effect. That explanation depended on his concept of light as consisting of unitary mono-directional particles concentrated in a laterally narrow body able to pass completely and freely through narrow slits and focus all of its action on objects as minute as a particular electron among others in a particular atom, those particles subsequently given the name "photons".

But now, having found that, except for narrowly collimated light in wave bursts from a laser, it is impossible for light to be of particle form and must be a form of wave with a wave front much more broad that that of Einstein's photon the issue arises "what is the correct mechanism of the photoelectric effect ?

How does incoming E-M radiation become absorbed by an atom's electron and excite it to being entirely free of the atom as in the photoelectric effect ?

First it is necessary to consider what it is that is absorbed. The light or E-M radiation available for absorption by and excitation of an electron is the same E-M radiation given off by an electron falling from a higher to a lower energy orbit. The radiation from any single outer-to-inner-orbit transition contains the energy, angular momentum and force that are exactly correct to cause an electron to execute the inverse transition.

However, the E-M wave from a single electron orbital transition must radiate outward from the transitioning electron. It is impossible for all of that radiation to encounter one other single electron. Rather, any single electron encounters a very small portion of that radiation's total wave front.

E-M radiation consists of a large number, a continuum, of waves propagating in the form of <u>individual half-cycle sinusoids of radiation</u>. A light beam or other radiation is a plethora of such <u>radiation bursts</u>, burst upon burst, side by side, in front, behind, overlapping, running together, and on and on. It is from among that immense number of very small parts of myriad electron inward transition radiations that that which an electron needs to perform an outward transition is found.

The wave front of a single such radiation burst, the output of a single electron's inward transition, disperses in space as it travels outward from its source. Only a quite small part of the total wave front of such an individual burst can encounter and interact with another particular electron. But, the radiation encountering a single particular electron is the sum of a very large number of such individual small parts of the radiation from individual electron inward transitions.

Out of the plethora of arriving half-cycle sinusoids, more precisely the plethora of small portions of their individual total wave fronts, the electron responds to an instantaneous sum of arriving waves that may or may not have the energy and momentum sufficient to free the electron from its atom - the photoelectric effect.

17

In a beam of light shined on a material, such as the light shined on a substance to yield the photoelectric effect, there is a very large number of radiative type interaction half-cycle sinusoid bursts propagating as waves. These constitute the beam of light. Some portion of them will coincide properly in frequency and timing and have sufficient collective amplitude at some electron to produce the observed photoelectric behavior.

"Sufficient collective amplitude" is the state in which the sum of the myriad minute parts of numerous transitions of the correct type adds up to being equal to or greater than the effect as if all of the radiation had passed intact from one such inward transitioning electron directly to the encountered electron (as the photon theory deemed it).

But, why is the energy magnitude dependent only on the frequency; how does wave amplitude enter into the process ? The waves inverse square disperse as *Propagated Outward Flow* does, so that the amplitude of each individual burst decreases steadily in inverse square manner from its value at the moment of the interaction that created it. The farther that an absorptive interaction is from the source of incoming radiation the greater the number of individual bursts, each contributing a small portion of the requisite amplitude, that are required to become the "sufficient collective amplitude" for an absorptive interaction to be able to take place.

The amount of energy naturally depends on frequency. The higher the frequency the more rapidly the E-M radiation oscillates. The E-M radiation carries the ability to cause corresponding change in motion in encountered charged particles. It requires more energy per time to make a rapid change than to make a gradual one. A shorter period (higher frequency) half-cycle sinusoid must contain directly proportional greater energy to produce the proportionally more rapid change.

CONCLUSION

The concept in general of the particle nature of light has been dominant in physics for over a century. That concept developed long before the invention of the laser, a device able to generate particle-like photons as treated in Section 9. Likewise, neglect of the wave nature of matter and focus on its particle conception has long dominated physics.

Having now found that it is impossible for normal light to be unitary mono-directional particles and, therefore it must be a form of wave, we now proceed to resolving the matter wave frequency problem and thus opening the wave nature of matter for analysis.

18

PART III -- THE WAVE NATURE OF MATTER

Having found that light cannot be particles the analysis now proceeds to showing how and why matter is wave in nature.

In order to correctly understand the nature of matter it is necessary to consider all of the applicable sources of information and data. There are two such sources:

- The behavior of matter in its various encountered circumstances, and
- The origin of matter – how and from what it came to be.

The behavior of matter has been thoroughly investigated over the years and is codified in what we may refer to as 20th Century physics. That is the starting point of this present work, the various "Laws", "particles", "forces" and so forth that are the current generally accepted understandings of how the material world functions physically.

Until the present the origin of matter, its source, has not been addressed and that omission has resulted in a major error in the understanding of the nature of matter – the incorrect solution to the problem of the wave nature of matter versus its particle nature.

- Solution to the Matter Wave Problem

- Atomic Electron Stable Orbits

- The Origin of Matter

- The Consequent Form of Matter

- The Outward Flow Propagated by All matter

- Relativistic Effects – The Propagated Waveform

- Matter Waves

21

SECTION 3

The Matter Wave Solution

The problem with matter waves was the failure to obtain a satisfactory matter wave frequency after DeBroglie developed the matter wave wavelength, a failure that resulted largely in abandonment of interest in matter waves. The solution to the problem is obtained from the relativistic calculation of kinetic energy.

EINSTEIN'S DERIVATION OF RELATIVISTIC KINETIC ENERGY

Kinetic energy, KE, is defined as the work done by the force, f, acting on the particle or object of mass, m, over the distance that the force acts, s. This quantity is calculated by integrating the action over differential distances.

(3-1)
$$KE = \int_0^s f \cdot ds \qquad \text{[Per above definition]}$$

$$= \int_0^s \frac{d(m \cdot v)}{dt} \cdot ds \qquad \text{[Newton's 2nd law]}$$

$$= \int_0^{(m \cdot v)} \frac{ds}{dt} \cdot d(m \cdot v) \qquad \text{[Rearrangement of form]}$$

$$= \int_0^{(m \cdot v)} v \cdot d(m \cdot v) \qquad [v = {ds}/{dt}]$$

$$= \int_0^v v \cdot d\left[\frac{m_r \cdot v}{\left[1 - \dfrac{v^2}{c^2}\right]^{\frac{1}{2}}} \right] \qquad \begin{array}{l} \text{[m is } m_r \text{ Lorentz} \\ \text{contracted by v.} \\ m_r \text{ is rest mass]} \end{array}$$

23

(3-1 continued)

$$= \frac{m_r \cdot v^2}{\left[1 - \frac{v^2}{c^2}\right]^{\frac{1}{2}}} - m_r \cdot \int_0^v \frac{v \cdot dv}{\left[1 - \frac{v^2}{c^2}\right]^{\frac{1}{2}}} \qquad \text{[Integration by parts]}$$

(3-2)

$$KE = \frac{m_r \cdot v^2}{\left[1 - \frac{v^2}{c^2}\right]^{\frac{1}{2}}} - m_r \cdot c^2 \cdot \left[1 - \frac{v^2}{c^2}\right]^{\frac{1}{2}} - m_r \cdot c^2 \qquad \text{[Integration of 2nd term]}$$

$$= \frac{m_r \cdot v^2 + m_r \cdot c^2 \cdot \left[1 - \frac{v^2}{c^2}\right]}{\left[1 - \frac{v^2}{c^2}\right]^{\frac{1}{2}}} - m_r \cdot c^2 \qquad \text{[Place 2nd term over 1st term denominator]}$$

$$= \frac{m_r \cdot v^2 + m_r \cdot c^2 - m_r \cdot v^2}{\left[1 - \frac{v^2}{c^2}\right]^{\frac{1}{2}}} - m_r \cdot c^2 \qquad \text{[Expand term with brackets]}$$

$$= \frac{m_r \cdot c^2}{\left[1 - \frac{v^2}{c^2}\right]^{\frac{1}{2}}} - m_r \cdot c^2 \qquad \text{[Simplify]}$$

(3-3) $\quad KE = m_v \cdot c^2 - m_r \cdot c^2 \qquad$ [m_v is total mass at $v > 0$

$\qquad\qquad\qquad\qquad\qquad\qquad m_r$ is total mass at $v = 0$

$\qquad\qquad\qquad\qquad\qquad\qquad m_v = m_r$ Lorentz transformed]

This result equation *(3-3)* states that:

{Kinetic Energy} = {Total Energy} - {Rest Energy}

or

{Total Energy} = {Kinetic Energy} + {Rest Energy}

The appearance in equation *(3-3)* that the energies are the product of the masses times c^2, the speed of light squared, was the origination of that concept, the famous Einstein's $E = m \cdot c^2$. The concept falls out naturally from applying the Lorentz transforms to the classical definition of kinetic energy. It is somewhat surprising that Einstein was the first to do that inasmuch as it was Lorentz who developed the Lorentz transforms and the Lorentz contractions.

ALTERNATIVE TREATMENT OF THE SAME DERIVATION

If in the above original derivation one proceeds onward differently from the first line of equation *(3-2)*, as below, a slightly different result is obtained.

(3-2 first line repeated)

$$KE = \frac{m_r \cdot v^2}{\left[1 - \frac{v^2}{c^2}\right]^{\frac{1}{2}}} - m_r \cdot c^2 \cdot \left[1 - \frac{v^2}{c^2}\right]^{\frac{1}{2}} - m_r \cdot c^2$$

(3-5)

$$KE + m_r \cdot c^2 = \frac{m_r \cdot v^2}{\left[1 - \frac{v^2}{c^2}\right]^{\frac{1}{2}}} - m_r \cdot c^2 \cdot \left[1 - \frac{v^2}{c^2}\right]^{\frac{1}{2}} \quad \text{[Move the "} - m_r \cdot c^2 \text{"]}$$

Considering and evaluating the three terms of equation *(3-5)*:

(3-6) $KE + m_r \cdot c^2$ = Kinetic plus rest energies

= Total Energy

= $m_v \cdot c^2$

(3-7) $\dfrac{m_r \cdot v^2}{\left[1 - \frac{v^2}{c^2}\right]^{\frac{1}{2}}}$ = A relativistically increased energy of motion which equals zero when v = 0.

= $m_v \cdot v^2$

(3-8) $m_r \cdot c^2 \cdot \left[1 - \frac{v^2}{c^2}\right]^{\frac{1}{2}}$ = A relativistically reduced rest energy which equals the at rest energy when v = 0

= Equation *(3-6)* – Equation *(3-7)*

= $m_v \cdot c^2 - m_v \cdot v^2$

the result is that equation *(3-5)* is equivalent to

(3-9) $\begin{bmatrix} \text{Total} \\ \text{Energy} \end{bmatrix}$ = $\begin{bmatrix} \text{Energy in} \\ \text{Kinetic Form} \end{bmatrix}$ + $\begin{bmatrix} \text{Energy in} \\ \text{Rest Form} \end{bmatrix}$

$m_v \cdot c^2$ = $m_v \cdot v^2$ + $m_v \cdot (c^2 - v^2)$

and (dividing the above energy equation by c^2 to obtain an equation in mass)

(3-10) $\begin{bmatrix} \text{Total} \\ \text{Mass} \end{bmatrix}$ = $\begin{bmatrix} \text{Mass in} \\ \text{Kinetic Form} \end{bmatrix}$ + $\begin{bmatrix} \text{Mass in} \\ \text{Rest Form} \end{bmatrix}$

m_v = $m_v \cdot v^2 / c^2$ + $m_v \cdot (1 - v^2/c^2)$

25

It is shown in Part IV, Section 6, that "in kinetic form" is real, and is an effect of the speed of light.

Why is the formulation for classical *Kinetic Energy* $KE = \frac{1}{2} \cdot m \cdot v^2$ but *Energy in Kinetic Form* is simply $m \cdot v^2$ without the $\frac{1}{2}$? When dealing with quite small velocities (v very small relative to c) the excursion of total energy above rest energy and the excursion of energy in rest form below rest energy are both essentially linear. In that case the portion above the rest case is essentially half of the total excursion above and below the rest case. The classical kinetic energy is then half, $\frac{1}{2} \cdot m \cdot v^2$, $\frac{1}{2}$ times the total energy in kinetic form, $m \cdot v^2$, for $[v/c]$ quite small.

APPLICATION TO THE PROBLEM OF THE MATTER WAVE

Thus the traditional view of kinetic energy as the energy increase due to motion may not be valid as a description of the processes taking place. Before the encountering of the relativistic change in mass with velocity the traditional view did not lead to problems in spite of its being an over-simplification.

Using mass- and energy-in-kinetic-form to obtain the frequency of the matter wave proceeds as follows.

(3-11)
$$f_{mw} = \frac{m_V \cdot v^2}{h}$$
[equation *(2-6)*, but using W_V, energy-in-kinetic-form, for W_k, kinetic energy]

Using this result for matter wave frequency and using the same relativistic mass, m_V, in equation *(2-5)* for the matter wavelength the velocity of the matter wave then is

(3-12)
$$v_{mw} = f_{mw} \cdot \lambda_{mw}$$

$$= \left[\frac{m_V \cdot v^2}{h} \right] \cdot \left[\frac{h}{m_V \cdot v} \right]$$

$$= v$$

and the wave is traveling with and as the particle.

APPLICATION TO THE ATOMIC ELECTRONS STABLE ORBITS

On that basis the wave aspect of matter is then established both experimentally (Davison and Germer and their successors) and theoretically (the above development). That gives new significance to the fact, observed at the time of Bohr's development of the relationship between atomic line spectra and atomic orbital structure, that the orbital lengths of the stable orbits of atomic electrons are an integer multiple of the orbiting electron's matter wave length.

The fact of the stable orbits has long been accepted without a specific reason, a specific operative cause, for those orbits and only those orbits being stable. The matter wave of the orbiting electron now provides an operative reason, as follows.

For the orbit to be stable it must be the same for each pass, pass after pass. If each pass includes exactly an integer number of the orbital electron's matter wave lengths then each pass is the same in that regard. But if, for example, the orbital path length contains only $9/10$ of a matter wave length, $9/10$ of the matter wave period, then the next pass will contain the missing $1/10$ of the matter wave length or wave period plus $8/10$ of the next, and so on. The matter wave being sinusoidal in form, the successive orbital passes will be all different.

It is this behavior which operatively causes the "stable orbits", and only those orbits, to be stable. It has nothing to do with angular momentum nor quantization of angular momentum. For the angular momentum hypothesis there is no underlying reason nor mechanism to produce stability or instability. The quantization of angular momentum concept is merely an invented defined condition, without operative cause, just as were the "stable orbits" it seeks to explain until their being here justified in terms of the operative matter wave behavior

The statement that the orbital electron's angular momentum is quantized, as in the following traditional equation

$(3\text{-}13)$ $\qquad m \cdot v \cdot R = n \cdot \dfrac{h}{2\pi}$ $\qquad\qquad\qquad [n = 1, 2, ...]$

is merely a mis-arrangement of

$(3\text{-}14)$ $\qquad 2\pi \cdot R = n \cdot \dfrac{h}{m \cdot v} = n \cdot \lambda_{mw}$ $\qquad\qquad [n = 1, 2, ...]$

a statement that the orbital path length, $2\pi \cdot R$, must be an integral number of matter wavelengths, $n \cdot \lambda_{mw}$, long. The latter statement has a clear, simple, operational reason for its necessity. The former statement is arbitrary and is justified only because it produces the correct result, even if without an underlying rational reason.

The assumption without any justification or support that the orbital electron's angular momentum is quantized is part of the early foundations of Quantum Mechanics along with Einstein's unjustified assumption that light is particulate in form for the photo-electric effect.

The Origin of Matter: Its Cause

INTRODUCTION

In order to correctly understand the nature of matter it is necessary to consider all of the applicable sources of information and data. There are two such sources:

- The behavior of matter in its various encountered circumstances, and

- The origin of matter – how and from what it came to be.

Causality or mechanism is apparent from observation and experience which show that every thing and every event has a cause, and that those causes are themselves the results of precedent causes, and *ad infinitum*. Defining and comprehending the causality or mechanism operating to produce any contended or proposed scientific truth is essential to authenticating or validating that truth.

Until the present the science community has addressed only the first of those two with regard to the nature of matter and the omission of the second has resulted in a major error in the understanding of the nature of matter – the incorrect solution to the problem of the wave nature of matter versus its particle nature.

HOW THE UNIVERSE'S MATTER CAME TO BE

We are confronted with an apparently insuperable problem. Before the universe there was nothing, absolute nothing. That is the starting point because it naturally occurs; it is the only starting point that requires no cause, no explanation nor justification for its existence. But, that starting point has two impediments to the universe, or anything, coming into existence from it. First is the problem of change from nothing to something without, at least initially, an infinite rate of change, which is impossible. Second is the problem of change from nothing to something without violating conservation, which must be maintained.

The analysis would appear to end at that point, end with the declaration that obviously there cannot be a universe and there is no universe. Except, of course, that we and the universe we inhabit clearly exist at least enough for us to investigate it. Therefore, a solution to the insuperable problem exists. That solution is as follows.

1 - THE PROBLEM OF INFINITE RATE OF CHANGE

To avoid a material infinity the rate of change at the moment of the change must have been finite. Rather than an instantaneous jump from nothing to something, no matter how small or "negligible" that something might have been, there had to be a gradual transition at a finite rate of change. Further, the rate of change of that rate of change, the change's second derivative, at that moment had to have been finite, and so on *ad infinitum* for all of the further derivatives.

That requirement means that the form of the change had to have been either a natural exponential or some form of sinusoid. That develops as follows, in which the sought form of the change will be the function $U(t)$ [the "U" for universe, of course].

To illustrate the problem consider the function

(4-1) $U(t) = 0$ $t < 0$
$U(t) = t^2$ $t = 0$ and $t > 0$

as a theoretical candidate for $U(t)$ at the beginning of the universe, which function is graphically depicted at the right.

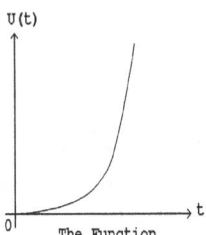

U(t)

The Function

Its first derivative, also depicted graphically to the right, is

(4-2) $\dfrac{dU(t)}{dt} = 0$ $t < 0$

$\dfrac{dU(t)}{dt} = 2 \cdot t$ $t > 0$

and is unstated for $t=0$ because $dU(t)/dt$ is not smooth there even though $U(t)$ "looks" smooth there.

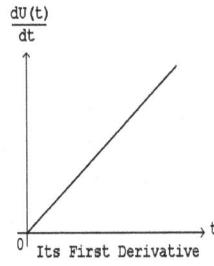

$\dfrac{dU(t)}{dt}$

Its First Derivative

Figure 4-1a

Now, the second derivative depicted graphically to the right

(4-3) $\dfrac{d^2U(t)}{dt^2} = 0$ $t < 0$

$\dfrac{d^2U(t)}{dt^2} = 2$ $t > 0$

is clearly discontinuous at $t=0$, the instant of the beginning of the universe, where it instantaneously jumps from 0 to 2 as depicted to the right.

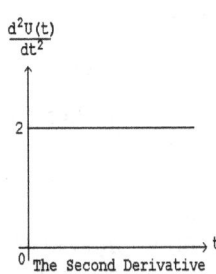

$\dfrac{d^2U(t)}{dt^2}$

2

The Second Derivative

Figure 4-1b

The third derivative, which is the rate of change of the second derivative must be infinite at $t=0$ to produce the instantaneous jump from 0 to 2. Clearly, that cannot have happened in the real universe. It is such a condition which is unacceptable in a candidate function for $U(t)$ at the beginning of the universe.

The only way to avoid that condition of an infinite derivative somewhere along the line of successive further derivatives is to have a function with an endless family of finite, non-zero derivatives; that is, some derivatives may be zero at $t=0$ but there must always be further non-zero higher derivatives, which requires that the functional form of every derivative must be non-zero.

One can conceive theoretically of the idea of a function for which all derivatives are non-zero and no two are alike (in a general sense analogous to the pattern of digits in an irrational number), but it is not likely that such a function can exist. In any case the more certain and more simple way to achieve all non-zero derivatives is a repeating derivative function, the two simplest examples of which are as below.

30

(4-4)   dU(t)

$$\frac{dU(t)}{dt} = \pm\ U(t) \quad \text{[First derivative = the original function]}$$

(4-5)   d²U(t)

$$\frac{d^2U(t)}{dt^2} = \pm\ U(t) \quad \text{[Second derivative = the original function]}$$

a. *Analysis of Repeating Derivative Functions*

Case (a): *Functions Satisfying Equation 4-4*

The function meeting this requirement is the natural exponential, ε^t.

(4-6)

$$\varepsilon^t = 1 + t + \frac{t^2}{2!} + \frac{t^3}{3!} + \cdots$$

Taking the first derivative

(4-7)

$$\frac{d[\varepsilon^t]}{dt} = 0 + 1 + \frac{2t}{2!} + \frac{3t^2}{3!} + \cdots$$

$$= 1 + t + \frac{t^2}{2!} + \frac{t^3}{3!} + \cdots = \varepsilon^t$$

so that the original function results as is required by equation *4-4*.

That is the prime case of a function that satisfies the requirement of all derivatives existing in functional form. In general those of this case are as equation *4-8*.

(4-8) $U(t) = A \cdot \varepsilon^t$

The function ε^t is not suitable for $U(t)$ at the beginning of the universe, however, because its value at $t=0$ is not zero. In fact it is zero only at $t = -\infty$. A function that might seem usable, however, would be

(4-9) $U(t) = 0 \qquad\qquad t < 0$ and $t = 0$

$U(t) = \varepsilon^t - 1 \qquad\quad t > 0$

$$= t + \frac{t^2}{2!} + \frac{t^3}{3!} + \cdots$$

which does have zero value at $t=0$ and otherwise meets the derivatives requirement sufficiently.

Cases (b) – (e): *Functions Satisfying (4-5)*

Turning to functions that meet the requirement that the second derivative equal the original function per equation *4-5* there are four such functions.

(4-10)

$$\text{Case (b):} \quad U(t) = 1 + \frac{t^2}{2!} + \frac{t^4}{4!} + \cdots$$

31

(4-11)

Case (c): $U(t) = 1 - \dfrac{t^2}{2!} + \dfrac{t^4}{4!} + \cdots$

(4-12)

Case (d): $U(t) = t + \dfrac{t^3}{3!} + \dfrac{t^5}{5!} + \cdots$

(4-13)

Case (e): $U(t) = t - \dfrac{t^3}{3!} + \dfrac{t^5}{5!} + \cdots$

These five candidate functions can be described and summarized as their exponential equivalents as in Figure 4-2, below.

Case	Function	Name of Function	Candidate U(t)
(a)	ε^t	Natural exponential	$\varepsilon^t - 1$
(b)	$\dfrac{\varepsilon^t + \varepsilon^{-t}}{2}$	Hyperbolic cosine	$\mathrm{Cosh}(t) - 1$
(c)	$\dfrac{\varepsilon^{i \cdot t} + \varepsilon^{-i \cdot t}}{2i}$	Cosine	$\mathrm{Cos}(t) - 1$
(d)	$\dfrac{\varepsilon^t - \varepsilon^{-t}}{2}$	Hyperbolic sine	$\mathrm{Sinh}(t)$
(e)	$\dfrac{\varepsilon^{i \cdot t} - \varepsilon^{-i \cdot t}}{2i}$	Sine	$\mathrm{Sin}(t)$

Figure 4-2

The relationships in the table can be verified by substitution using the formula for ε^t as given in equation *4-6*, above. Cases *(b)* and *(c)* have the same problem that case *(a)* had, that the value of $U(t)$ is not zero at *t=0*. Just as with case *(a)*, they would appear to become satisfactory if a constant, *1*, is subtracted from each of them.

These candidates all satisfactorily meet the requirement for a continuous family of derivatives so that the kind of unacceptable problem as encountered in the example of $U(t) = t^2$ at the beginning of this discussion is avoided. That is, all derivatives are finite. But, there are other requirements that the successful $U(t)$ function must meet.

b. *Using the Remaining Criteria to Select U(t)*

Two other criteria must be met by the successful candidate function or functions:

- the function must not be open-ended, that is it cannot ever have an infinite amplitude, and

- the function must smoothly match the $U(t)=0$ condition at *t=0*.

The first criterion eliminates cases *(a)*, *(b)* and *(d)* each of which goes to an infinite value of $U(t)$. To satisfy the second criterion the tangent to $U(t)$ at $t=0$ must be identical to the tangent to the function for $t < 0$, which is the horizontal t-*axis*. The condition is satisfied if the first derivative of $U(t)$ equals *zero* at $t=0$. Only cases *(b)* and *(c)* meet that requirement.

Therefore, the resulting form of $U(t)$, the only acceptable form, the only one that meets all of the requirements, is case *(c)*,

$$(4\text{-}14) \quad U(t) = [\text{Cos}(t) - 1] \qquad\qquad t > 0 \text{ and } t = 0$$
$$U(t) = 0 \qquad\qquad\qquad\qquad t < 0.$$

which is identical in form to the more usual and convenient equation *4-15*.

$$(4\text{-}15) \quad U(t) = U_0 \cdot [1 - \text{Cos}(2\pi \cdot f \cdot t)]$$

in which an amplitude parameter, U_0, and a frequency parameter, f, have been added.

That the only possible form for the manner in which the universe began is a sinusoidal oscillatory form would seem to be very appropriate. Oscillations, waves, are ubiquitous in our universe from oceans, violin strings and pendulums to sound, light and electron orbits. That statement can also be validly inverted: Oscillations and waves are ubiquitous in our universe because the universe began from an initial such oscillatory form.

Every oscillation that we know in nature exhibits, and the very theory of oscillations in the abstract requires, that the oscillation consist of two aspects storing and exchanging the energy of the oscillation back and forth by means of a "flow". (With one aspect varying in oscillatory fashion then when that aspect decreases there must be some "place" for its energy to go, a place in which it is stored until it reappears in that aspect when it increases again. It cannot completely disappear or be lost because the oscillation would die. That "place" is the oscillation's second aspect and it obviously must vary in a manner related to the first aspect's variation, but with its energy storage in opposite phase.

A pendulum, for example, oscillates by the motion (flow) of its swinging mass between peak height in the gravitational field (potential energy) at each end of the swing and peak speed of motion (kinetic energy) at the mid-point between the ends of the swing. Then, what is the "flow" of the original oscillation at the start of the universe ? We do not know and likely will never know but we can give it a name, *Medium*, and we can investigate its characteristics and nature.

Such was the oscillation at the beginning of the universe except that at the first half cycle the energy was in only one form increasing from zero to its maximum. Then the second form began, similarly from zero to maximum, receiving and storing the energy of the first form as that gradually decreased in the second half cycle.

2 - THE PROBLEM OF CONSERVATION – "SOMETHING FROM NOTHING"

At this point, that is the universe having started from absolute nothing as an oscillation having the form of equation *4-15*, the maintaining of conservation, the avoiding of getting something from nothing, clearly could only happen in one manner:

There simultaneously had to have arisen an identical-in-form but opposite-in-amplitude oscillation so that the pair balanced out to the original net nothing, as in equation *4-16*.

(4-16) $U(t) = \pm U_0 \cdot [1 - Cos(2\pi \cdot f \cdot t)]$

There is no other way that violating the assured principle of conservation could have been avoided. The universe exists. It had to come into being from a prior nothing. That had to happen while avoiding an infinity of rate of change. Conservation had to be maintained. Therefore the universe began with the oscillation of equation *4-16*.

3. THE PROBLEM: WHY THAT OSCILLATION BEGAN AND WHAT IT WAS

a. Why That Beginning happened

A duration is the period of time that a particular state or set of conditions persists. The duration is terminated by a change, which change also initiates a new duration. In the universe change is ubiquitous. It is the constant and continuous stream of change that makes durations mensurable. Before the beginning of the universe a duration was in process even though it was not mensurable. The beginning of the universe was the first change ever and it terminated the original primal duration of absolute nothing.

The probability of the happening of such an event is extremely small. But the event was / is not impossible. Furthermore, in the absence of that event occurring there was an extremely large duration of opportunity in which that extremely small probability could operate. In the absence of the beginning the original duration would have been infinite and that infinite opportunity operated on by minute, but non-zero, probability results in absolute certainty. The beginning of the universe could not avoid eventually happening.

b. What That Beginning Oscillation Was

The starting point is the assumption that, when the primal nothing changed as a probabilistically inevitable interruption of what would otherwise have been an infinite duration of the primal nothing, the simplest or minimum conservation-maintaining interruption that could occur is what occurred. There are two reasons for this. Occam's Razor, calls for the simplest hypothesis as the most likely. More importantly, or perhaps the same thing, if an essentially spontaneous and extremely low probability event is to occur solely as an interruption of the duration of an otherwise absolute nothing, then very little interrupting event is needed; the barest minimum of something is sufficient to interrupt, to be a change in absolute nothing. There is no call, no reason for anything more. So, while the interruption could have been otherwise, it was probably as simple and minimum as possible.

Size or amount of time are of no meaning here because there is nothing to which they can be compared or by which they can be measured. Whatever amount of change occurred is what occurred. Whatever time it took, or went on for, whatever its oscillatory frequency was, is what happened. Twice as much or half as much have no meaning.

The following conclusions about the initial oscillatory $\pm U_0 \cdot [1 - Cos(2\pi \cdot f \cdot t)]$ form can now be reasonably obtained:

- clearly the universe of today must be an on-going evolved consequence of its beginning, of the initial oscillatory form;

34

- the frequency, f, of the sinusoidal oscillation was, and is, very large; and

- the nature of the change is one of concentration or density of the something that is oscillating.

- the oscillation was spherical, radially outward in all directions from its origin, because there was nothing to constrain it otherwise.

The frequency would have to be either very large or very small -- high enough so that it is not detected or noticed by us in every day life or so low that it appears to us as no change at all in our experience.

It has already been noted that the fact that the only possible form for the manner in which the universe began is a sinusoidal oscillatory form is very appropriate because oscillations, waves, are ubiquitous in our universe from oceans, violin strings and pendulums to sound, light and electron orbits. And it has been noted that that statement can be validly inverted: oscillations and waves are ubiquitous in our universe because the universe began from an initial such oscillatory form.

If the frequency of the initial oscillation were so small that it appears to us as no change at all it would completely eliminate oscillations playing any significant part in the behavior of the universe as we know it. Therefore, the frequency must have been very large, so rapid compared to our perception that we do not notice the oscillation at all.

The change can hardly be one of gross size if it is going on right now at high frequency as has just been concluded. One can conceive of the fundamental "substance", the "something" of the universe flashing into and out of existence from a zero to a maximum density or concentration in an oscillatory fashion at a rate so high that we neither detect nor notice it at all. But, it is not possible to entertain a concept of reality flashing from zero to full size, a size that includes ourselves and our environment, in such a fashion.

Actually, the reality that we know is not "flashing into and out of existence ...." Our reality is more the oscillation itself than what is oscillating and the continuing oscillation is our steady, constant reality.

> Thus the interruption that gave us our universe was the starting of an *oscillation* that was *spherical*, present to us at a very high frequency and of $\pm U_0 \cdot [1 - Cos(2\pi \cdot f \cdot t)]$ form, of the density, as the variation will be hereafter referred to, of the *Medium*, as what it is that is oscillating will be hereafter referred to.

All of the discussion so far must apply to the "negative" oscillation, $-U(t)$, exactly as to the "positive" oscillation $+U(t)$ because the exact same reasoning as for $+U(t)$ applies to $-U(t)$ and, after all, they are not distinguishable in the discussion. The terms "+" and "-" are merely terms of convenience for two equal form opposite magnitude unknown things. We probably tend to think of our universe as the "+", but that is meaningless and irrelevant. There can be no objective designation of $+U(t)$ and $-U(t)$, no way to identify one versus the other. Both had to appear and our universe cannot avoid being the evolved result of both.

The universe that we know and exist in is the combined integrated result of both $+U(t)$ and $-U(t)$. The "+" and "-" electric charges of our universe [in both matter as

35

for example in protons and electrons and in anti-matter as for example in negaprotons and positrons] must derive from that aspect of the beginning. (It is interesting to observe, also, that our universe being the integrated result of an initial beginning and its opposite relates to (presumably is the underlying cause of) the dialectical nature of reality, the ying and yang of oriental philosophy.)

The question of what the *Medium* is can only be answered in terms of its characteristics, what it does and how. Its characteristics are:

- a continuous entity, not a mass of "particles" nor anything having parts,

- simple and uniform throughout,

- of minimum tangibility or substantiality, not unlike the actuality of what we designate as "field" [electric, gravitational, etc].

4. THE PROBLEM: WHY DID THE EFFECTS OF EQUATION 4-16 NOT PROMPTLY CANCEL AND ON-GOING ABSOLUTE NOTHING RESUME ?

This is resolved in detail in Appendix A, *Why No Immediate Mutual Annihilation*. Briefly, the initial structure was so unstable that it promptly exploded in that which we refer to as the "Big Bang" before total annihilation could occur.

5. THE PROBLEM: IT HAS BEEN THOUGHT THAT THE UNIVERSE HAD TO START AT A POINT. HOW COULD A POINT DELIVER A WHOLE UNIVERSE?

The sole reason for positing a point origin was to avoid an initial infinite rate of change. The gradualness of the $[1 - Cosine]$ form resolves the problem of avoiding an infinite rate of change so that a point origin is no longer required.

The Big Bang "event horizon" problem and its relation to the development of variety in the universe has led to the hypothesis that there was an initial brief period of extremely rapid expansion called "inflation". That hypothesis has no supporting cause nor mechanism except its role in meeting the "event horizon" problem.

But with the need for a point origin eliminated the origin can have started per equation $4-16$ at any size. There was no un-accounted-for period of "inflation". From estimates calculated of the number of particles in today's universe it has been determined that the initial, at the very first instant, the already "inflated"–size universe began. It was a highly concentrated volume of all of the mass and energy of the universe of about $40,000$ km radius [the Big Bang's "core" see Section 6.

That size is in terms of today's sizes. For that event specific size is meaningless because there was nothing else to compare it to.

The development of the oscillatory wave form of matter based on its origin from the primal absolute nothing continues in the following Section 5.

The Behavior of Matter: Its Form

Section 4, *The Origin of Matter: Its Cause* resolved the origin of the matter of the universe as follows.

> The universe exists. It had to come into being from a prior nothing. That had to happen while avoiding an infinite rate of change. Conservation had to be maintained. *Ergo* equation *4-16* repeated below.

(4-16) $U(t) = \pm U_0 \cdot [1 - \cos(2\pi \cdot f \cdot t)]$

Thus the hypothesis is that the interruption that started our universe, the interruption of what would otherwise have been an infinite duration of the primordial absolute nothing, an interruption because an essentially infinite amount of opportunity operated on a non-zero though minute probability, was the starting of a matched pair of spherical oscillations:

- Present to us at a very high frequency,

- Of the general *[1 - Cosine]* form, and

- Together equal to the original nothing because of having
  matching amplitudes $+U_0$ and $-U_0$.

That analysis yielded an initial event, the origin oscillations, as in Figure 5-1. [All of the unavoidably planar depictions of the spherical oscillations are of the spherical phenomenon, interpretable as a radial versus time depiction.]

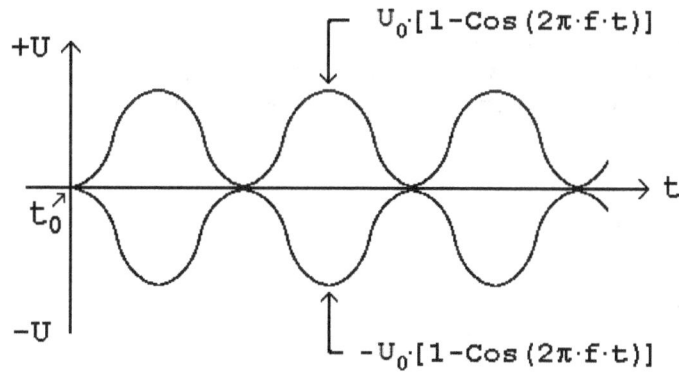

Figure 5-1

HOW THE ORIGINAL OSCILLATIONS BECAME THE UNIVERSE

Examination of the waveform of Figure 5-1 reveals two problems. One, that it is an immediate mutual annihilation, will be dealt with shortly below. Of concern now is that an infinite rate of change still remains; the envelope of the oscillation has an infinite rate of change at $t=t_0$ as can be seen in Figure 5-2, below, which displays the envelope.

Viewed in a mathematical or graphical sense without any consideration of the physical reality represented, the envelope discontinuity at $t=t_0$ is not a difficulty. The only quantity that actually exists and is varying is the overall $U(t)$. The envelope is merely our perception of a characteristic of the waveform. The actual varying quantity, per Figure 5-1, has no discontinuity at $t=t_0$

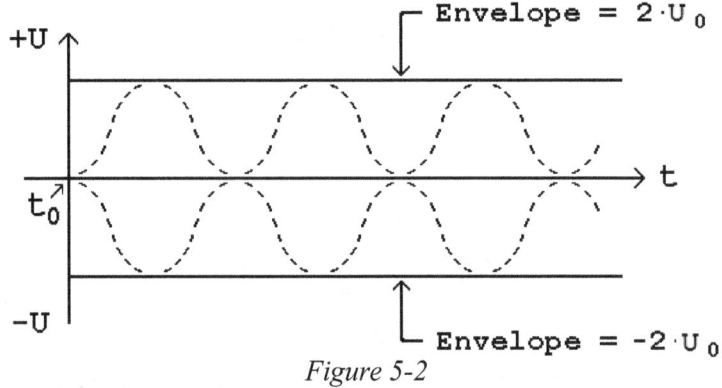

Figure 5-2

However, looking at the situation in a physical sense rather than purely mathematically, such oscillations as depicted in Figure 5-1 are all that there is to account for the effects which we call *energy, mass* and *charge*. Therefore, this *energy / mass / charge / oscillation* is something other than nothing. It is a physical reality that did not exist prior to the Origin. It can no more leap from zero to a finite non-zero amount than could the original $U(t)$ so leap.

That infinite rate of change in the amount of *energy / mass /charge* at $t=t_0$ is no more acceptable than was the infinite rate of change encountered in the original analysis of the beginning and it must be corrected by the same kind of reasoning as was then pursued: the envelope, also, had to originate as a *[1 - Cosine]* form of oscillation, which is the only form that avoids an infinite rate of change and matches the requirements of the situation.

That original envelope oscillation was at a lesser frequency than the original wave by the definition of a waveform envelope. If it were at a greater frequency then the roles (envelope and wave) would be reversed. If it were at the same frequency it would not act as an envelope and the infinity problem would remain. If we designate the envelope frequency as f_{env} and the frequency of the wave oscillation within the envelope as f_{wve} then the envelope would be of the following form.

(5-1) $\quad U_{env} = [1 - Cos(2\pi \cdot f_{env} \cdot t)]$

The wave is, as before, of the form

(5-2) $\quad U_{wve} = \pm U_0 \cdot [1 - Cos(2\pi \cdot f_{wve} \cdot t)]$

and the envelope modulating the wave is then

(5-3) $\quad U(t) = [U_{env}] \cdot [U_{wve}]$

$\qquad\qquad = \pm U_0 \cdot [1 - Cos(2\pi \cdot f_{env} \cdot t)] \cdot [1 - Cos(2\pi \cdot f_{wve} \cdot t)].$

That waveform appears in Figure 5-3.

However, the form of $U(t)$ of equation *5-3* and Figure 5-3 still does not resolve the problem of an infinite rate of change at t_0. The *[1 - Cosine]* envelope is itself an oscillation that begins at t_0 with a sudden step from zero to its full amplitude. Figure 5-3 shows the first *2* cycles of the envelope oscillation, which if only the envelope is considered, is a simple oscillation

40

at the envelope frequency, even though visually, in the Figure, it is only the trace of the peaks of the overall complex oscillation.

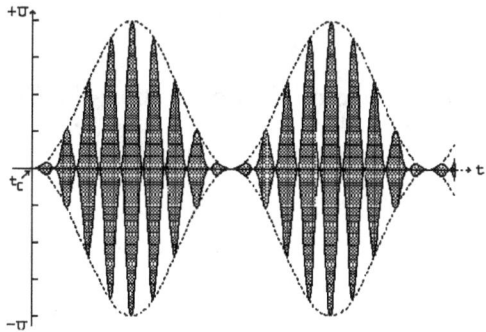

Figure 5-3

It is *energy / mass / charge* that begins suddenly in its full amount at t_0 just as, in Figure 5-1, the oscillation of equation 5-1 begins at t_0. Therefore, it is again necessary to introduce an envelope of *[1 - Cosine]* form to prevent the infinite rate of change at t_0 in the prior envelope. That correction will in turn require still another such correction and so *ad infinitum*. An (apparently at this point) infinite string of envelopes thus results as a necessity of the situation.

The resulting $U(t)$ then is

(5-4)

$$U(t) = \pm U_0 \cdot \prod_{i=1}^{i=\infty} \left[[1 - \text{Cos}(2\pi \cdot f_{env_i} \cdot t)] \right] \cdot \ \cdots$$

$$\cdots \ \cdot \left[[1 - \text{Cos}(2\pi \cdot f_{wve} \cdot t)] \right]$$

where the $\prod$ symbol (a large π, Greek "p")
means the product of the indicated factors.

While an envelope frequency must be less than the frequency of the wave that it modulates so that the various f_{env} must be less than f_{wve}, each successive envelope may be at the same frequency, as the prior. The reason is as follows.

If each envelope frequency must be different then each must be at least slightly smaller than the prior. With an infinite set of envelopes and only the frequency range from slightly less than that of the wave down to slightly above zero being available each successive envelope could only be at an infinitesimally lower frequency than its predecessor in any case. Infinitesimally less is essentially the same as identical.

Then how did other than an infinite string of envelopes come about ?

Each additional envelope factor in equation 5-4 results in a higher frequency content in the overall expression. That is, as each envelope is added the expansion of the exponentiated cosines expression into a sum of individual frequency cosine terms becomes longer and acquires higher frequency terms. But, the oscillation could not have had an actual component at infinite frequency. The real universe original $U(t)$ had an enormous set of envelopes but not an infinite set; they were "cut off" at some point.

41

The *Medium* of these oscillations being the only reality and, therefore, being what sets the limit on the speed of light with which we are familiar, the *Medium* also sets a limit on the highest frequency / lowest wavelength waves that can propagate. As a result the series of envelopes, of factors in equation *5-4*, was limited to some finite but quite large amount as developed in Appendix B, *The Limitation of the Original Envelopes*).

This yields a revised *U(t)*, the original oscillation, the Cosmic Egg, as equation *5-5*, below. N_0 is the number of envelopes, all at the same frequency, f_{env}.

(5-5) $$U(t) = \pm U_0 \cdot \left[1 - \cos\left[2 \cdot \pi \cdot f_{env} \cdot t\right]\right]^{N_0} \cdot \left[1 - \cos\left[2 \cdot \pi \cdot f_{wve} \cdot t\right]\right]$$

The waveform *[1 - Cos(x)]^n* converges to an increasingly narrower peak as *n* increases, Figure 5-4, below. For very large *n*, that is very large N_0 of equation *5-5*, the converging of the waveform into a single narrow peak proceeds to a momentary "spike" per cycle. Figure 5-5, below, shows the appearance of the waveform for extremely large *n*, that is for *n* = N_0 - what the waveform of the original "Cosmic Egg", the start of our universe, "looked like". (N_0 is found further below to be about 10^{84}.)

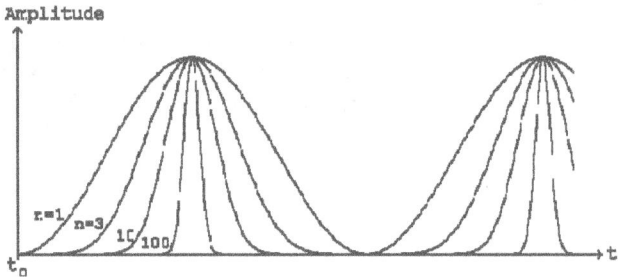

Figure 5-4 *[1 - Cos(x)]^n For n = 1, 3, 10, 100*

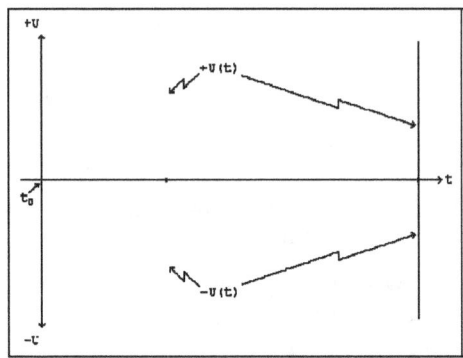

Figure 5-5
The U(t) "Cosmic Egg" Waveform

This discussion of *U(t)*, the original oscillation the start of which was the start of the universe, has dealt so far only with the problems of the Origin, the problems of the transition from nothing to something. The something was, of course, the first instant of the entire universe. As such it must have contained in itself all of the *mass / energy / positive and negative charge* of the universe.

Figures 5-1, 5-3, and 5-5 all indicate that the original pair of oscillations, *+U* and *-U*, should have immediately mutually annihilated, canceled out, reverted to the primal nothing. But, clearly that did not happen. The only explanation of that not happening is that each was unstable, so unstable that they exploded more immediately than they were able to mutually annihilate. They immediately proceeded to an immense explosion of energy and pieces of their oscillation, the event now called the "Big Bang". See Appendix A, *Why No Immediate Mutual Annihilation.*

In terms of the $U(t)$ as depicted in Figure 5-5, the so immediate explosive decay undoubtedly occurred after only a minute portion, an infinitesimal portion, of the very first cycle had passed. It had to have been long before the first "spike". In that sense the initial event was very small, tenuous, hardly more than nothing because the instantaneous amplitude of $U(t)$ at that moment (the height of the curve above zero at that moment long before the first "spike") was also infinitesimal. It was hardly more than, essentially zero.

In that sense, the way that the universe started at all becomes a little more comprehensible. To avoid an infinite rate of change there was essentially almost no difference between "nothing", on-going absolute nothing, and the first infinitesimal moment of the original $U(t)$, the original oscillation.

Yet, it contained the entire universe.

THE FORM OF MATTER AS GENERATED BY THE "BIG BANG"

What did that "Cosmic Egg" explode into ? It could only explode into pieces of what it was made of, pieces of *[1 – cosine]* form spherical oscillations, pieces like equation *4-16*, above.

Each oscillation is three-dimensional, thus spherical, because three dimensions is the minimum number that can involve space part of which is not its own boundary.

But, what did the "Cosmic Egg" explode into ? It primarily exploded into what we know our universe to mainly consist of: myriad protons - Hydrogen atom nuclei, and myriad electrons - maintaining overall charge neutrality with the protons, and the antimatter forms of both, negaprotons and positrons – maintaining conservation.

[Those might also be expected to have mutually annihilated but did not. Their survival rather than annihilation is analyzed in full in Appendix A, *Why No Immediate Mutual Annihilation.* Suffice it here to observe that each product piece was initially ejected radially outward at extreme velocity and energy, on paths slightly diverging, such that initially annihilations could not occur.]

Then, what was the nature, the form of those product pieces that the "Cosmic Egg" exploded into ? Because of the two frequencies of $U(t)$, f_{wve} and f_{env}, and that the explosion source was of two equal but opposite polarities, $+U_0$ and $-U_0$, the "Big Bang" resulted in myriad pieces of four different forms of *[1 – cosine]* form spherical oscillations , equations *5-6*.

(5-6)
$$U_{Form\ 1}(t) = +U_c \cdot [1 - Cos(2\pi \cdot f_{wve} \cdot t)] \quad \text{the proton}$$

$$U_{Form\ 2}(t) = -U_c \cdot [1 - Cos(2\pi \cdot f_{env} \cdot t)] \quad \text{the electron}$$

$$U_{Form\ 3}(t) = -U_c \cdot [1 - Cos(2\pi \cdot f_{wve} \cdot t)] \quad \text{the anti-proton}$$

$$U_{Form\ 4}(t) = +U_c \cdot [1 - Cos(2\pi \cdot f_{env} \cdot t)] \quad \text{the anti-electron}$$

Each of those has a specific value of its mass. Per the data provided by NIST, the National Institute of Standards and Technology those masses are:

(5-6a) ■ the proton and the antiproton $m_p = 1.672\ 621\ 898 \cdot 10^{-27}\ kg$

■ the electron and the anti-electron $m_e = 9.109\ 383\ 56 \cdot 10^{-31}\ kg.$

Using the mass-energy relationship, $m \cdot c^2 = h \cdot f$ the frequency, f, of those particles can be calculated. Those frequencies are:

(5-6b)   ■ the proton and anti-proton: $f_{wve} = 2.268,731,818 \cdot 10^{23}$ hz

■ the electron and anti-electron: $f_{env} = 1.235,589,965 \cdot 10^{20}$ hz.

Finally, the mass of those four fundamental particles having now been resolved, their electric charge remains. They all have the same magnitude of their oscillation, $|U_c|$, which by default is the magnitude of their electric charge. [U_c is the particle oscillation amplitude per equation 5-6. U_0 is the original pre-explosion oscillation amplitude.] The magnitude of the oscillation is in two opposite polarities; therefore clearly, where q is the fundamental electric charge per NIST, then:

(5-7) $q = 1.602,176,621 \times 10^{-19}$ c

$+U_c = +q$

$-U_c = -q$

Judging by its result, the "Cosmic Egg" was not unlike an immense atom, a very unstable immense atom [as are all of the atomic species of atomic number exceeding 83 which the cosmic egg would have immensely exceeded]. Its "Big Bang" was a kind of explosive nuclear radioactive decay ultimately ending in the myriad stable elements of today's Periodic Table plus those unstable with half lives long enough to be in detectable quantities today. Such decays follow a chain:

- From a heavy and complex composition,

- To various multiple less heavy less complex product pieces,

. . . . .

- Until they arrive at many multiple stable forms.

Some of those "multiple less heavy less complex product pieces" having long half lives are present to us still today still decaying as what we term "radioactive" species.

The vast majority of those resulting stable forms are the protons and electrons of the material world and their anti-particles. They are of the equation 5-6 form spherical oscillation, and will be referred to as *Spherical-Centers-of-Oscillation*.

SUMMARY

The form of matter is not that of the "particles" of classical modern physic's Standard Model. Rather the form of matter is *Spherical-Centers-of-Oscillation*, spherical oscillations of *[1 - Cosine]* form as equation 5-6.

[For the action of matter in the various laws of physics see R. Ellman, "On the Nature of Matter" 2018, The-Origin Foundation, Inc., available at Amazon.com.]

44

The Outward Flow Propagated by all Matter

Section 5 demonstrated that the fundamental particles of atoms and therefore of matter in general are *Spherical-Centers-of-Oscillation* in the form of equation *5-6*. The present Section 6 develops the details of the structure and behavior of those fundamental wave-in nature-particles.

THE FLOW FROM THE SPHERICAL-CENTERS-OF-OSCILLATION

The Particle "Core"

Consider a small individual particle such as a proton. Where a_{grav} is the gravitational acceleration, G is the Newtonian constant of gravitation, and d is the separation distance of the source and the acted-on Newton's law of gravitation expressed in terms of the masses, m_{source} and $m_{acted-on}$, and with both sides of the equation divided by $m_{acted-on}$ is, of course,

$$(6\text{-}1) \qquad a_{grav} = G \cdot \left[\frac{m_{source}}{d^2} \right]$$

However, mass and energy are equivalent, so that [using c = light speed and h = Planck's constant] a mass, m, is proportional to an oscillation frequency, f, that is characteristic of that mass. That is

$$(6\text{-}2) \qquad m \cdot c^2 = h \cdot f \quad \text{or} \quad f = [\,c^2/h\,] \cdot m$$

so that the m_{source} of   (6-1 has a corresponding equivalent frequency, f_{source}.

That being the case, the gravitational acceleration, a_{grav}, can be expressed in terms of that frequency as the change, Δv, in the velocity, v, of the attracted mass per time period, T_{source}, of the oscillation at the corresponding frequency, f_{source}, as follows.

$$(6\text{-}3) \qquad a_{grav} = \Delta v\,/\,T_{source} = \Delta v \cdot f_{source}$$

It can then be reasoned using equation *6-3* = equation *6-1* as follows .

$$(6\text{-}4) \qquad a_{grav} = \Delta v \cdot f_{source} = G \cdot \left[\frac{m_{source}}{d^2} \right]$$

Equation *6-5)*, below, is obtained by using that frequency is proportional to mass. With f_p and m_p as the proton frequency and mass then $f_{source} = [m_{source}\,/\,m_p] \cdot f_p$.

$$(6\text{-}5) \qquad \Delta v \cdot \left[\frac{m_{source}}{m_p} \right] \cdot f_p = G \cdot \left[\frac{m_{source}}{d^2} \right]$$

Rearranging and canceling m_{source} on both sides of the equation,

$$(6\text{-}6) \qquad \Delta v = \frac{G \cdot m_p}{d^2 \cdot f_p} \quad \text{per cycle of } f_{source}.$$

Then substituting, per equation 6-2), $m_p = [\, h \cdot f_p \,] / c^2$,

$$(6\text{-}7) \qquad \Delta v = \left[\frac{G}{d^2 \cdot f_p} \right] \cdot \left[\frac{h \cdot f_p}{c^2} \right]$$

$$= \frac{G \cdot h}{d^2 \cdot c^2} \quad \text{per cycle of } f_{source}.$$

The Planck Length, l_P, is defined as

$$(6\text{-}8) \qquad l_P \equiv \left[\frac{h \cdot G}{2\pi \cdot c^3} \right]^{\frac{1}{2}} \quad \text{so that} \quad G = \left[\frac{2\pi \cdot c^3 \cdot l_P{}^2}{h} \right]$$

Substituting G as a function of the Planck Length from equation 6-8 into G as it is in equation 6-7), the following is obtained.

$$(6\text{-}9) \qquad \Delta v = \left[\frac{2\pi \cdot c^3 \cdot l_P{}^2}{h} \right] \cdot \left[\frac{h}{d^2 \cdot c^2} \right]$$

$$= c \cdot \frac{2\pi \cdot l_P{}^2}{d^2} \quad \text{per cycle of } f_{source}.$$

This means that at distance $d = \sqrt{2\pi} \cdot l_P$ from the center of the source, attracting mass, the acceleration, Δv, per cycle of that attracting mass's equivalent frequency, f_{source}, is equal to the full speed of light, c, the most that it is possible to be. In other words, at that [quite close] distance from the source mass the maximum possible gravitational acceleration occurs. That is the significance, the physical meaning, of l_P or, rather, of $\sqrt{2\pi} \cdot l_P$.

The physical significance of $\sqrt{2\pi} \cdot l_P$ is that it sets a limit on the minimum separation distance of particles and therefore that a "core" of that radius is at the center of fundamental particles. That is, equation 6-9 clearly implies that it is not possible for a particle to be approached closer than that distance.

That physical significance of $\sqrt{2\pi} \cdot l_P$, is so fundamental to particle structure, that it more truly represents a fundamental constant than does l_P. For that reason that length should replace l_P as a fundamental constant of nature as follows.

$(6\text{-}10)$ <u>The fundamental distance constant, δ</u>

$$\delta^2 \equiv 2\pi \cdot l_P{}^2$$

$$\delta = 4.051,34 \times 10^{-35} \text{ meters}$$

The Particle Core's Propagated Outward Flow

Each gravitationally attract**ing** *Spherical-Center-of-Oscillation* must tell each gravitationally attract**ed** particle its "message": the direction from the attract**ed** particle back to the attract**ing** one and the magnitude of the attract**ing** particle's gravitational attraction. That task is assigned by contemporary physics' theory to a *gravitational field*, a vector field that is an assignment of a direction of action and its magnitude to each point in a region of space.

However, that designation of the field, while facilitating the description of the action fails to explain the cause, the mechanism of the field and thus fails to explain or account for the action at issue. It also fails to account for the time delay due to the limitation of the speed of light that must exist between a change at the attract**ing** particle and its effect at the attract**ed** particle.

Something flowing is required, something flowing at the speed of light, continuously, carrying the direction and magnitude information, spherically outward, from every gravitating *Spherical-Center-of-Oscillation* to every other *Spherical-Center-of-Oscillation*.

Furthermore, the necessity for gravitation that an oscillation and its frequency are closely involved in the effect [equation *6-9*] and therefore in what is communicated by the flow, means that the <u>flow itself is oscillatory</u> corresponding to and generated by its oscillatory source, the *Spherical-Center-of-Oscillation*.

For such a flow to persist there must be a supply of that outward flowing substance in every particle. And, for that flow to have persisted the billions of years since the "Big Bang" that "supply" must be an extremely concentrated reservoir of that which flows outward [concentrated relative to the outward flow].

Having now just determined:
- That δ sets a limit on the minimum separation distance in gravitational interactions and therefore that a "core" of that radius is at the center of fundamental particles, and
- That an extremely concentrated reservoir supply of that which is flowing outward from those particles is required at the center of all particles to support the billions of years of their outward flow;

Therefore:
- The reservoir is the spherical "core" of radius δ at the center of all particles;
- That it is impenetrable is because of its immense density concentration [billions of years worth of flow of the flow substance [*Medium*] in the minute ($\delta = 4.05134 \times 10^{-35}$ *meters* radius spherical core) of every particle], and.
- The *Spherical-Center-of-Oscillation* is a spherical oscillation of that immensely concentrated flow substance.

Then, what "contains" that core's supply or why doesn't it all just quickly "slosh" out and be gone ? The answer is that it is trying to do just that, to "slosh" out, as hard as it can. It cannot help propagating outward because it has no container, no physical boundary. But it can only propagate outward at the limiting rate determined by its surface area, $4 \cdot \pi \cdot \delta^2$ and the fastest speed possible for flow, the speed of light, c. Thus is the *Propagated Outward Flow* of what we term *medium*.

The Speed of the Flow – The Speed of Light

Every oscillation that we know in nature exhibits, and the very theory of oscillations in the abstract requires, that the oscillation consist of two aspects of the substance which is oscillating [e.g. pendulum position and velocity or electric potential and current] storing and exchanging back and forth the energy of the oscillation. With one aspect varying in oscillatory fashion then when that aspect decreases there must be some "place" for its energy to go, a place in which it is stored until it reappears in that aspect when it increases again. It cannot completely disappear or be lost because the oscillation would die. That "place" is the oscillation's second aspect and it obviously must vary in a manner related to the first aspect's variation with its energy storage in opposite phase.

The matter of the universe is largely a mass of particles each a spherical *[1 - Cosine]* form oscillation propagating outward.

Like electric inductance and capacitance determining the speed of propagation along a transmission line, μ_0 and ε_0 determine the speed of the *[1 - Cosine]* form oscillation propagation in "free space" by setting the two aspects of the oscillation in which they are involved, the aspects between which the oscillation energy exchanges back and forth.

But, when the original oscillation came into existence it did so in absolute nothing. There was no "free space" with μ_0 and ε_0. There was nothing but the original oscillation. And, after the immediate explosion into all of the particles of the universe, each of those particles was sending its *Propagated Outward Flow* into nothing, into emptiness.

Where did the *Propagated Outward Flow*'s μ_0 and ε_0 come from? The only thing they could have come from was the source of the oscillation. There is no other possible source because everything else was absolute nothing, "the zero of existence". The μ_0 and ε_0 are inherent in the substance of the oscillation, which means, μ_0 and ε_0 are also inherent in the outward propagation. Each particle's *Propagated Outward Flow* contains its own μ_0 and ε_0.

The form of matter is not that of the "particles" of classical modern physic's Standard Model. Rather the form of matter is:

- *Spherical-Centers-of-Oscillation*, spherical oscillations of *[1 - Cosine]* form;

- Propagating spherically outward a continuous oscillatory *Propagated Outward Flow* of *Medium* in *[1 - Cosine]* form, according to its source *Spherical-Center-of-Oscillation* magnitude, sign, and frequency;

- The speed of the *Propagated Outward Flow*, c, being set by the net μ_0 and ε_0 in the *Medium* being propagated;

(6-11)
$$c = \frac{1}{\sqrt{\mu \cdot \varepsilon}}$$

SUMMARY

The *Spherical-Center-of-Oscillation* consists of a central "core", a spherical volume of radius $\delta = 4.051,34 \times 10^{-35}$ meters that consists entirely of a high density concentration of the oscillating *Medium*, which propagates outward at an extremely low rate as restricted by the surface area of the "core" and the radial outward speed of flow of the propagated *Medium*, the speed of light, c.

That *Propagated Outward Flow* is a communication from every particle in the universe to every other – an intra-particle network of communication.

The next section, Section 7, addresses the issue of what is the information so communicated.

Relativistic Effects on the Flow Waveform

THE PROBLEM

As found in Section 6, a *Spherical-Center-of-Oscillation* naturally sends a *Propagated Outward Flow* of *Medium* uniformly radially outward in all directions from itself at velocity c, the speed of light. As presented on page 51 at "The Speed of the Flow – The Speed of Light" the speed of that flow is set by the μ_0 and ε_0 of the flowing *Medium* to the exact value of c by virtue of their controlling the cyclical alternating exchange of the oscillation between the two forms in which it exists.

When the center is not in motion that presents no problem, but with the *Spherical-Center-of-Oscillation* moving in some direction the center's motion and its propagation are in conflict. In the direction of motion the velocity of the center, v, tends to add to the natural value of the speed, c, of propagation of the *Propagated Outward Flow* and in the opposite direction it tends to subtract. But, the speed of the flow is fixed; set at c by μ_0 and ε_0.

That conflict forces an adjustment of the oscillation of the *Spherical-Center-of-Oscillation* to modify the propagation speed of its *Propagated Outward Flow*.

THE SPHERICAL-CENTER-OF-OSCILLATION AT CONSTANT VELOCITY

The treatment is of the *Spherical-Center-of-Oscillation* at constant velocity because that is the most direct and simple case of motion, and at constant velocity one cannot detect absolute motion. That is, one can say that there is a relative difference of velocity between two systems at constant velocity in one of which the observer is located, but the observer cannot say which system is moving and which, if any, is at rest.

To describe the behavior of the center its propagation will be modeled resolved into three components: forward, rearward, and sideward relative to the direction of the center's velocity, as depicted in Figure 7-1. [In the figure the "up", "down", "left" and "right" are all "sideward".] These orthogonal components represent the propagated wave in all directions. The wave in any particular direction is the "resultant" of that directions' projection on the forward or rearward component (whichever is at a nearer angle) and on the sideward component. (The "resultant" is the hypotenuse of the right triangle having the projection components as its other two sides.)

53

Where λ_r and f_r are the wave length and frequency of the *Propagated Outward Flow* when its center is at rest [absolute "rest" relative to its propagation] then propagation of waves is the same in all directions at speed $c = \lambda_r \cdot f_r$.

A Center-of-Oscillation at Rest

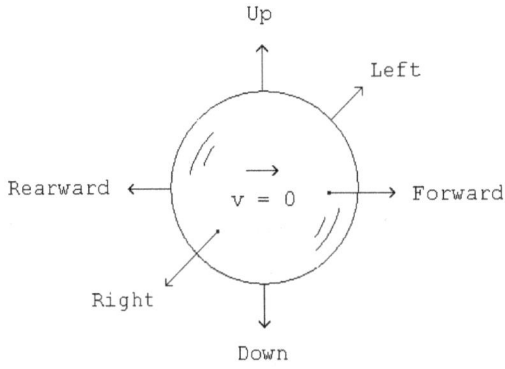

Figure 7-1

As described earlier the speed of flow of centers' propagation is fixed at c by the μ_0 and ε_0 of the flowing *Medium*. The center moving at velocity v would find (in the forward direction) its freshly emitted propagation "thrown" forward at speed $[c + v]$ interfering with the flow just ahead of it at speed c and conflicting with the μ_0 and ε_0 of the *Medium*. It finds the propagated wave not moving out of the way at the needed $[c + v]$ in time for the next cycle as set by the at-rest frequency of the center. The result is an imperative to reduce the center frequency ["delay" the next cycle] by the factor $[1 - v/c]$. That "interfering" and "conflicting" tends to force on the center a change in its oscillation, a reduction by the factor $[1 - v/c]$. That is, with the center moving forward at v,

(7-1) Propagated Speed would become $c \cdot [1 - v/c] = (c - v)$

   Flow speed = propagated speed + v = $(c - v) + v = c$

In the rearward direction the opposite is the case, an imperative to increase the center frequency by the factor $[1 + v/c]$. But, the *Spherical-Center-of-Oscillation* can only oscillate at one specific frequency at a time. It cannot both increase and decrease its oscillation frequency at the same time. It responds by adopting a compromise change in frequency, the geometric mean of the two conflicting factors as in equation 7-1.

With subscript v meaning "at velocity v" the center's oscillation frequency decreases and its oscillation wavelength correspondingly increases, the product still being c.

(7-2)

$$f_v = f_r \cdot \left[1 - \frac{v^2}{c^2}\right]^{\frac{1}{2}}$$ [Center frequency decreases]

$$\lambda_v = \lambda_r \cdot \frac{1}{\left[1 - \frac{v^2}{c^2}\right]^{\frac{1}{2}}}$$ [Center wavelength increases]

$$\lambda_v \cdot f_v = \lambda_r \cdot f_r = c$$ [Wave velocity still at c]

54

While the center can oscillate at only one frequency, it can propagate at different wavelengths in different directions. To maintain propagated wave velocity at c in the direction of center motion the wave must be actually propagated forward by the center at $c-v$ relative to the center itself so that the wave velocity relative to at rest is the propagated velocity, c, plus the center velocity, v, that is $(c-v)+v = c$.

To propagate forward at $[c-v]$ while maintaining the frequency at f_v requires that the wavelength change to a smaller value, λ_{fwd}. Likewise, rearward the wave must be actually propagated by the center at $[c+v]$ relative to the center with a greater wavelength, λ_{rwd}. Those adjusted propagation wavelengths are as follows.

(7-3)

$$\lambda_{fwd} = \frac{c-v}{f_v} = \frac{c\left[1-\dfrac{v}{c}\right]}{f_r\left[1-\dfrac{v^2}{c^2}\right]^{\frac{1}{2}}} = \lambda_r \cdot \frac{\left[1-\dfrac{v}{c}\right]^{\frac{1}{2}}}{\left[1+\dfrac{v}{c}\right]^{\frac{1}{2}}} = \lambda_r\left[\frac{c-v}{c+v}\right]^{\frac{1}{2}}$$

$$f_{fwd} = \frac{c}{\lambda_{fwd}} = f_r\left[\frac{c+v}{c-v}\right]^{\frac{1}{2}}$$

$$\lambda_{rwd} = \lambda_r\left[\frac{c+v}{c-v}\right]^{\frac{1}{2}}$$

$$f_{rwd} = f_r\left[\frac{c-v}{c+v}\right]^{\frac{1}{2}}$$

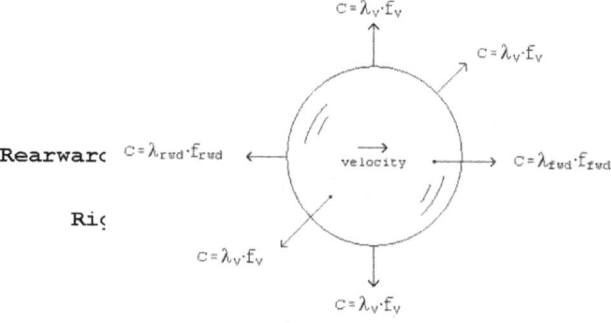

Figure 7-2, The Wave as Propagated by the Center at Velocity v (relative to the center)

Figure 7-3, The Above Propagation (as Observed from At-Rest)

As the center "sees" it, per the above Figure 7-2, it is oscillating at f_v, with the forward and rearward wavelengths adjusted for the velocity so that the wave travels in each direction at speed c. As "at-rest" would "see" it, per Figure 7-3, above, the center appears to propagate different forward and rearward frequencies, f_{fwd} and f_{rwd}.

Thus the field of propagated waves is traveling at c in all directions as observed by the center that is in motion and doing the propagating and as observed from at-rest.

THE EFFECT OF VELOCITY ON MASS

With the oscillation frequency corresponding to the rest mass of the particle it represents the development so far of decreasing oscillation frequency. Equation *7-2*, demonstrates a decrease in rest mass due to the *Spherical-Center-of-Oscillation's* velocity. That is more properly referred to as a decrease in that part of the mass effect due to the overall frequency of oscillation of the center, to be referred to as "mass in rest form", m_r'.

(7-4)

$$m'_r = m_r \frac{f_v}{f_r} = m_r \left[1 - \frac{v^2}{c^2} \right]^{\frac{1}{2}}$$

Newton's Second Law and it as restated by inversion are:

$$Force = Mass \cdot Acceleration$$

$$Acceleration\ Resulting = Force\ Applied \times {}^1/_{Mass}$$

which translates in terms of the waves of *Propagated Outward Flows* and *Spherical-Centers-of-Oscillations* into

$$\begin{bmatrix} Acceleration \\ Resulting \end{bmatrix} = \begin{bmatrix} Wave \\ Impulse \end{bmatrix} \cdot \begin{bmatrix} Responsiveness \\ of\ the\ Center \end{bmatrix}$$

or, more succinctly,

$$Acceleration = Wave \times Responsiveness.$$

Of the total wave traveling outward from a source *Spherical-Center-of-Oscillation*, the only part that interacts with another *Spherical-Center-of-Oscillation* is the part intercepted by that encountered center. The *Spherical-Center-of-Oscillation* intercepting the larger portion of incoming wave receives the greater impulse, the greater momentum change. Thus center responsiveness depends on the encountered center's cross-section target for interception of *Propagated Outward Flow* waves and total mass depends inversely on that.

Per equation *7-4* the particle's rest mass decreases, however, overall the total mass increases because the effects so far have reduced the cross-section target for interception of *Propagated Outward Flow*. From the forward or the rearward point of view the center's cross-section is proportional to the area of the circle of radius λ_v, the sideward direction per equation *7-2*, repeated below.

(7-2)

$$\lambda_v = \lambda_r \cdot \frac{1}{\left[1 - \frac{v^2}{c^2} \right]^{\frac{1}{2}}}$$

56

That means that relative to the center's rest mass, m_r, the overall mass at velocity, m_v, is

(7-5)

$$m_v = m'_r \left[\frac{\lambda_v}{\lambda_r} \right]^2 = m_r \left[1 - \frac{v^2}{c^2} \right]^{\frac{1}{2}} \cdot \left[\frac{1}{\left[1 - \frac{v^2}{c^2} \right]^{\frac{1}{2}}} \right]^2$$

$$= m_r \cdot \frac{1}{\left[1 - \frac{v^2}{c^2} \right]^{\frac{1}{2}}}$$

From the sideward point of view the cross-section is no longer a circle, however. In the forward direction the at-rest circle's radius has become λ_{fwd} instead of λ_v and in the rearward direction λ_{rwd} instead of λ_v.

(7-6)

$$\lambda_{fwd} = \frac{c - v}{f_v} = \frac{c \left[1 - \frac{v}{c} \right]}{f_v} = \frac{\left[1 - \frac{v}{c} \right]}{\lambda_v} \qquad \text{therefore} \qquad \frac{\lambda_{fwd}}{\lambda_v} = \left[1 - \frac{v}{c} \right]$$

(7-7)

$$\lambda_{rwd} = \frac{c + v}{f_v} = \frac{c \left[1 + \frac{v}{c} \right]}{f_v} = \frac{\left[1 + \frac{v}{c} \right]}{\lambda_v} \qquad \text{therefore} \qquad \frac{\lambda_{rwd}}{\lambda_v} = \left[1 + \frac{v}{c} \right]$$

The product of the change factors, equations 7-6 and 7-7, is $[1 - v^2/c^2]$, a reduction of cross-section, the same amount of increase in mass as equation 7-5. Thus in all directions the effect of velocity is an increase in mass per equation 7-5.

THE CENTER OF OSCILLATION "AT REST" AND "IN MOTION"

In motion at a constant velocity, v, the *Spherical-Center-of-Oscillation* experiences the asymmetrical distortions of equation 7-3 and figures 7-2 and 7-3. The distortions indicate the motion and the motion enhanced energy of the center. At rest, in the absence of motion the center is spherically symmetrical.

Thus the rest mass and rest energy correspond to the spherically symmetrical portion of the center's oscillation [the only portion if v = 0] and they are "mass in rest form" and "energy in rest form". The overall distorted portion corresponds to the total "mass in kinetic form" and "energy in kinetic form" of the center.

Those distortions of the Propagated Outward Flow mean that that flow conveys information about the location, motion direction, velocity, mass and energy of its source *Spherical-Center-of-Oscillation*. Inasmuch as that flow is radially outward from every matter particle it constitutes a speed of light comprehensive communication among all of the particles of the universe

THE LORENTZ CONTRACTIONS, LENGTH AND TIME

Logic requires of the overall universe that in all frames of reference at constant velocities relative to each other [*i.e.* inertial frames]:

- The equations describing the laws of physics have the same form, and

- The universal constants appearing in those equations be the same,

This is called the Principle of Invariance, and means that the speed of light, c, a universal constant, is the same in all inertial frames, which appears to conflict with our instinctive assumption that the speed of light should vary with the speed of the light's source.

That logic combined with experiments showing that the speed of light actually is of the same value independent of whatever inertial frame, required the development of the Lorentz Transformations to account for the constancy of the speed of light. The transformations are coordinate transformations between two inertial frames. The Lorentz contractions are the related change in the fundamental quantities: mass, length, and time.

This brings up the point that, contrary to Einstein, there is an absolute frame of reference, an "at rest" frame. Einstein contended that there was not because he thought that an absolute frame could have different physical laws and constants. But, rather the opposite is the case. The absolute prime frame of reference is why the Principle of Invariance is valid. That is why it is required of the overall universe that in all frames of reference at constant velocity relative to each other [*i.e.* inertial frames]:

- The equations describing the laws of physics have the same form, and

- The universal constants appearing in those equations are the same,

The speed of light, c, a universal constant is the same everywhere.

They are all part of the same one overall absolute frame of reference, the rest frame. The rest frame is not special in that its laws and constants are not different. It is special in that all other frames are relative to it. It is simply the actual frame of the "Big Bang"

When the *Spherical-Center-of-Oscillation*'s oscillation is perfectly spherically symmetric then the center's velocity is zero and it is completely at rest. That is the universe's absolute frame to which all motion and all other frames are relative.

The Lorentz Contractions

The Lorentz Contractions are as follows.

(7-8)

$$L = L_r \cdot \left[1 - \frac{v^2}{c^2} \right]^{\frac{1}{2}}$$

[Observed Length in the Direction of motion shortens.]

$$f = f_r \cdot \left[1 - \frac{v^2}{c^2} \right]^{\frac{1}{2}}$$

[Observed frequency slows.]

$$t = t_r \cdot \frac{1}{\left[1 - \frac{v^2}{c^2} \right]^{\frac{1}{2}}}$$

[Observed time periods length, Time passes more slowly.]

$$m = m_r \cdot \frac{1}{\left[1 - \frac{v^2}{c^2} \right]^{\frac{1}{2}}}$$

[Observed mass increases.]

58

Time and frequency are reciprocals of each other and the above equation *7-2* decrease in center frequency with velocity validates the *f* and *t* Lorentz Transforms. [The increasing λ_r to λ_v of that equation is compensating for the frequency decrease to keep the sideward propagation speed at c. Sideward is not the direction of v so the Lorentz Contraction does not apply to that λ.]

The equation *7-5* overall increase in center mass with velocity validates the mass, *m,* Lorentz Transform. Remaining to be validated is the length, *L* contraction. The λ_{fwd} and λ_{rwd} contraction equations *7-6* and *7-7* are a *Spherical-Center-of-Oscillation* length contraction in the velocity direction, a Lorentz Contraction.

However on the macroscopic scale it is necessary to investigate two centers and the distance between them in order to develop a velocity-caused contraction of length in matter. In bulk matter composed of multiple particles, atoms and their components, the spacing of the atoms depends on the balance of the various electrostatic forces acting as a result of the centers-of-oscillation, protons and electrons, of which the matter atoms are composed.

Considering just two *Spherical-Centers-of-Oscillation* at rest in a fixed position relative to each other, the effect of their moving jointly at velocity *v* in the direction of the line joining them should be a Lorentz Contraction to closer spacing of the two centers by the Lorentz Contraction factor.

The position of each of the two centers is the balance of all of the forces acting on the centers, an equilibrium position. If the velocity is to change the distance between the two centers then the force acting between the two centers must change so that a new closer equilibrium spacing exists and determines the new distance between the two centers. For the centers to need to be closer in order to re-establish equilibrium the effective charge of each of the centers must be decreased by the velocity.

In other words, for the Coulomb force between the two centers

(7-9)
$$F = \frac{Q_1 \cdot Q_2}{4\pi \cdot R^2}$$

to be unchanged even though *R* is reduced by the Lorentz Contraction by the factor

(7-10)
$$\frac{R_{vel}}{R_{rst}} = \left[1 - \frac{v^2}{c^2}\right]^{\frac{1}{2}}$$

so that R^2 is changed by the factor

$$\frac{R^2_{vel}}{R^2_{rst}} = \left[1 - \frac{v^2}{c^2}\right]$$

then $Q_1 \cdot Q_2$ must be so reduced by the same factor as is R^2.

But, that is exactly the case. It has already been shown by equation *7-3* that the forward wave propagation speed is reduced by the factor *[1-^{v}/c]* to *c'* = *c-v* and that the rearward wave propagation speed is analogously changed by the factor *[1+^{v}/c]* to *c''* = c+v.

59

The charge Q corresponds to the impulse that the wave can deliver which depends directly on the above propagation speeds. The Q of the trailing center "looking" forward is reduced by the reduction of its c to $c' = c - v$, a factor of $[1-v/c]$.

Similarly the charge, Q, of the leading center "looking" backward is increased by the increase of its c to $c'' = c + v$, a factor of $[1+v/c]$.

Therefore, the equation 7-9 $Q_1 \cdot Q_2$ is reduced by the product of the two factors which is $[1 - v^2/c^2]$, which matches the Lorentz Contraction of R^2 and therefore of R and validates the length, L, Lorentz Contraction.

COMMENT

In this Section 7 we have, on page 55, development from fundamentals of the mechanism of the *Propagated Outward Flow* and the speed of light.

We have, on pages 61-62 development from fundamentals of the mechanism of the Lorentz Contractions.

In the pages 55-59 we have development from fundamentals of the reasons, the mechanism, that it is impossible for matter to have a velocity equal to or greater than the speed of light.

These developments make the point that every physics phenomenon, every material event, has a physics, a material, mechanism which causes and controls it. A phenomenon cannot be considered understood until its mechanism has come to be understood.

Further the validity of any contended phenomenon or behavior for which there is no developed mechanism is in question.

These observations apply specifically to the various "quantum" behaviors contended by Quantum Mechanics. For those the terminology "Mechanics" does not refer to mechanisms but only to contended phenomena.

For more information on the *Propagated Outward Flow*, specifically how it causes Coulomb's Law [electrostatic effects], Ampere's Law [magnetic effects] and Newton's Laws [gravitation and motion] see the References page ix.

Matter Waves

MATTER WAVES AND SPHERICAL CENTERS OF OSCILLATION

The matter wave traveling right along with the particle is like a kind of standing wave relative to the particle. A standing wave can be thought of as the sum result of two waves traveling in opposite directions through each other. If the frequencies and wavelengths are different then their interaction produces a new frequency called a "beat". The development of the beat is as follows.

The two waves are

(8-1)   Wave #1 $= A \cdot Sin(2\pi f_1 t)$

Wave #2 $= A \cdot Sin(2\pi f_2 t)$

and the sum is

(8-2) $\mathrm{WaveSum} = A \cdot Sin[2\pi f_1 t] + A \cdot Sin[2\pi f_2 t]$

which by using a trigonometric equivalence can be arranged as

$$\mathrm{WaveSum} = 2A \cdot Sin\left[2\pi \frac{f_1 + f_2}{2} t\right] \cdot Cos\left[2\pi \frac{f_1 - f_2}{2}\right]$$

The cosine term frequency $\frac{1}{2} \cdot [f1 - f2]$ difference, is smaller than the sine term sum $\frac{1}{2} \cdot [f1 + f2]$. If the expression is viewed as the higher frequency sine portion with the rest of the expression being the amplitude, as in equation 6-8, then

(8-3)

$$\mathrm{WaveSum} = \left[2A \cdot Cos\left[2\pi \frac{f_1 - f_2}{2} t\right]\right] \cdot Sin\left[2\pi \frac{f_1 + f_2}{2} t\right]$$

$$= [\,\mathrm{Varying\ Amplitude}\,] \cdot Sin\left[2\pi \frac{f_1 + f_2}{2} t\right]$$

The wave form appears as in Figure 8-1, below.

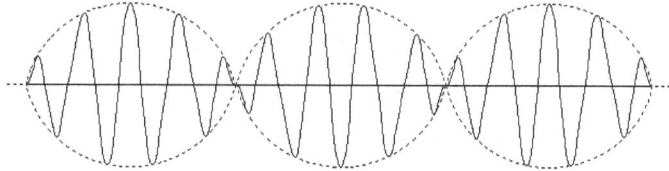

Figure 8-1

The solid-line curve in Figure 8-1 is the overall wave form. The dotted line, the *envelope*, is the varying amplitude. The overall wave form exhibits in the varying amplitude a periodic variation called the *beat*. The beat is real, not merely an appearance. For example two sound tones heard simultaneously produce an audible beat that one can hear. It is by listening to the beat that one tunes a piano or other musical instrument.

63

Matter waves are the beat that results from the *Spherical-Center-of-Oscillation*'s forward and rearward oscillations interacting with each other. This develops as follows. For a center in motion at velocity v.

(8-4)
$$\lambda_{fwd} = \lambda_v \cdot (1 - v/c) \qquad f_{fwd} = c/\lambda_{fwd}$$
$$\lambda_{rwd} = \lambda_v \cdot (1 + v/c) \qquad f_{rwd} = c/\lambda_{rwd}$$

The beat frequency, using the "Varying Amplitude" portion of equation *6-8*, substituting f_{fwd} for f_1 and f_{rwd} for f_2, and then using equation *6-9*, is

(8-5)
$$f_{beat} = \frac{1}{2}\left[f_{fwd} - f_{rwd}\right] = \frac{1}{2}\left[\frac{c}{\lambda_v\left[1-\frac{v}{c}\right]} - \frac{c}{\lambda_v\left[1+\frac{v}{c}\right]}\right]$$

$$= \frac{c}{2\cdot\lambda_v}\cdot\left[\frac{\left[1+\frac{v}{c}\right]-\left[1-\frac{v}{c}\right]}{\left[1-\frac{v}{c}\right]^2}\right] = \frac{v}{\lambda_v}\cdot\left[\frac{1}{\left[1-\frac{v}{c}\right]^2}\right]$$

$$\lambda_{beat} = \frac{c}{f_{beat}} = \lambda_v \cdot \frac{c}{v}\left[1-\frac{v^2}{c^2}\right]$$

Substitute Eqn $4-2$
$$\lambda_v = \lambda_r \cdot \frac{1}{\left[1-\frac{v^2}{c^2}\right]^{\frac{1}{2}}}$$

$$= \left[\lambda_r \frac{1}{\left[1-\frac{v^2}{c^2}\right]^{\frac{1}{2}}}\right]\frac{c}{v}\left[1-\frac{v^2}{c^2}\right]$$

Substitute per :
$$m\cdot c^2 = h\cdot f = h\cdot\frac{c}{\lambda} \rightarrow \lambda_r = \frac{h}{m_r\cdot c}$$

$$= \left[\frac{h}{m_r\cdot c}\right]\frac{c}{v}\cdot\left[\left[1-\frac{v^2}{c^2}\right]^{\frac{1}{2}}\right]$$

Substitute per Eqn $4-6$
$$m_v = m_r \cdot \frac{1}{\left[1-\frac{v^2}{c^2}\right]^{\frac{1}{2}}}$$

$$= \frac{h}{m_v\cdot v}$$

which is the matter wavelength as previously obtained per equation *6-1* (in which the mass must be relativistic mass, m_v, of course). Thus matter waves are the beat that results from the *Spherical-Center-of-Oscillation*'s forward and rearward oscillations interacting with each other.

A moving center-of-oscillation as "seen" by an external observer appears as the waves propagated by the center in his direction appear. But, if one could, somehow, actually "see" the center itself pulsating as it does, the situation would be different. The interaction of the forward and rearward oscillations, which produce a beat at the matter wave frequency, are real. The effect is as follows (repeating the form of equations *6-6* through *6-8*, which were for any general oscillation, but now using the oscillations of a center-of-oscillation in motion).

(8-6)   Forward Wave $= A\cdot\left[1+Sin(2\pi f_1 t)\right]$

Rearward Wave $= A\cdot\left[1+Sin(2\pi f_2 t)\right]$
```
[Note: 1 - cos(x) ≡ 1 + cos(180° - x)
                 ≡ 1 + sin[90°-(180° - x)]
                 ≡ 1 + sin(x - 90°)
       and the 90° phase is irrelevant, of course.]
```
and the sum is

(8-7)   WaveSum $= A\cdot\left[2+Sin\left[2\pi f_1 t\right]+Sin\left[2\pi f_2 t\right]\right]$

Which again by using a trigonometric equivalence can be arranged as

$$\text{WaveSum} = 2A + 2A \cdot \text{Sin}\left[2\pi\frac{f_1 + f_2}{2}t\right] \cdot \text{Cos}\left[2\pi\frac{f_1 - f_2}{2}\right]$$

The cosine term is at a lesser frequency than the sine term. If the expression for the wave sum is viewed as the (higher frequency) sine portion with the rest of the expression being the amplitude, as in equation 6-13, then

(8-8)

$$\text{WaveSum} = 2A \cdot \left[1 + \text{Cos}\left[2\pi\frac{f_1 - f_2}{2}t\right]\right] \cdot \text{Sin}\left[2\pi\frac{f_1 + f_2}{2}t\right]$$

$$= 2A \cdot \left[\begin{array}{l}1 + \text{cosine form of} \\ \text{Varying Amplitude}\end{array}\right] \cdot \text{Sin}\left[2\pi\frac{f_1 + f_2}{2}t\right]$$

In the case of a *Spherical-Center-of-Oscillation* $f_1 = f_{fwd}$ and $f_2 = f_{rwd}$. Likewise, A is U_C, the center average amplitude, the oscillation being of the form $U_C \cdot [1 - Cos(2\pi \cdot f \cdot t)]$.

The wave form appears as in Figure 8-2, below, for the forward-rearward interaction and the matter wave beat of the center's pulsation as it would be "seen" from the side relative to its direction of motion. Within the matter wave envelope is the center's spherical oscillation modified by the matter wave beat.

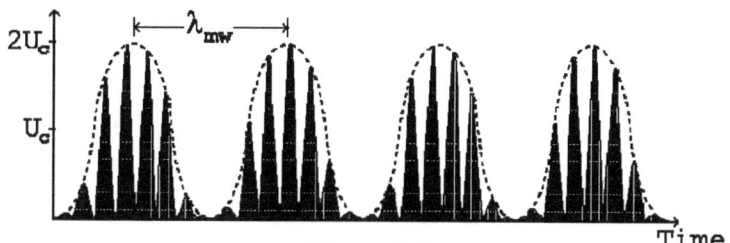

Figure 8-2
The Forward-Rearward Pulsation of a Center in Motion
Which is the Matter Wave

65

PART IV -- RESOLUTION OF THE "SPOOKINESS" OF QUANTUM MECHANICS

- *The Wave Nature of Matter and Light*
  Quantum Mechanics Mechanism

- *Resolving the Problem of Realism*

- *Resolving the Problem of Locality*

-*Resolving the Problem of Entanglement*

The Nature of Matter and Light
and Resolution of the Problem of Realism

THE PROBLEM

The mechanism of something is the process by which that thing takes place or is brought about. It is a cause of that thing's actions or effects. Quantum Mechanics is in denial of the dependency of effects on prior causes and of the principles of Realism and Locality. That state of denial stems from numerous validated experiments demonstrating various quantum effects conjoined with a total lack of identified causes and mechanisms for the observed behavior.

Such is the overriding problem of Quantum Mechanics, a problem that led Einstein to declare certain quantum behaviors as "spooky".

Furthermore, Quantum Mechanics appears to apply only to the physically very small (the "micro") leaving classical mechanics dealing with the rest of material reality (the "macro"), but with no reason or mechanism for that restriction.

The solution to the problems of Quantum mechanics develops from the wave nature of matter and of light: the oscillatory nature of matter particles' *Spherical-Centers-of-Oscillation* with their *Propagated Outward Flow* developed in Sections 5 and 6 and the dual particle - wave nature of the photon.

For Light, The Photon

Since 1960, when the first laser was made and operated, there have been two different forms of light. The foundations of Quantum Mechanics theory were developed in about 1920. Consequently, all experiments bearing on or cited with regard to Quantum Mechanics effects and for the foundations of Quantum Mechanics theory done before 1960 were experiments done with "natural light" because they preceded the advent of the laser.

"*Natural Light*" The primary source of light is the transition of an orbital electron of an atom from a "higher" "stable" orbit inward to a less high stable orbit. In *Section 2* it was found that it is impossible for such light to be unitary mono-directional particles and that it is a form of wave. Such light waves are generally of broad wave fronts and tend to spread out in space. They are not spatially limited to a narrow particulate-photon-like lateral dimension able fully to pass through a narrow opening. Per Section, 2 photons of such light are half-cycle sinusoidal waves in bursts of total energy of $W = h \cdot f$.

"*Coherent Light*" An important but much less ubiquitous form of light is that generated by a laser. A laser is a device that emits light through optical amplification based on the stimulated emission of electromagnetic radiation. The stimulated emission initially produces the above primary source kind of light but the amplification process results in the light becoming coherent. Coherent light is a beam of photons, particle-like light waves, again half-cycle sinusoidal waves in bursts of energy $W = h \cdot f$, that all have the same frequency and waveform. Only a beam of coherent laser light is able to largely resist spreading and diffusing.

For Matter Particles

In *Section 5* it was found that the form of matter is *Spherical-Centers-of-Oscillation*, spherical oscillations of *[1 - Cosine]* form. In *Section 6* it was found that The *Spherical-Center-of-Oscillation* consists of a central "core", a spherical volume of radius $\delta = 4.051,34 \times 10^{-35}$ meters that consists entirely of a high density concentration of the oscillating substance, *Medium*.

What "contains" that core's supply or why doesn't it all just quickly "slosh" out and be gone ? The answer is that it is trying to do just that, to "slosh" out, as hard as it can. It cannot help propagating outward because it has no container, no physical boundary. But it can only propagate outward at the limiting rate determined by its surface area, $4 \cdot \pi \cdot \delta^2$ and the fastest speed possible for flow, the speed of light, *c*. It is that flow which mediates the separation distance inherent in the Coulomb Effect, magnetic effects and Gravitation.

In *Section 7* it was found that the forward and rearward wave propagation, as well as the sideward, of a moving *Spherical-Center-of-Oscillation* are different and vary with the velocity of the particle. Those differences carry the information about the state of the particle including its direction, velocity, energy, frequency and mass.

In *Section 8* it was found that the matter wave of particles is a valid actual wave phenomenon that results from the 'beat' of the moving *Spherical-Center-of-Oscillation's* forward wave with it's rearward.

The intent of this work is not to dispute the numerous Quantum Mechanics experiments conducted and the physical results obtained. Nor is it to criticize attempts to obtain useful physical applications of that behavior such as for example quantum computers.

Rather, the problem of Quantum Mechanics is in the interpretations, the meanings associated with those experiments. It is widely recognized that there is dispute in the science world as to the correct interpretation. The most widely accepted interpretation of the various Quantum Mechanics phenomena is that called "The Copenhagen Interpretation" so named because of its development and advocacy by Neils Bohr and Werner Heisenburg working in the Danish City of Copenhagen.

For an effect, a behavior of nature, to validly demonstrably exist but with no explanation of how or by what mechanism it so exists is simply not acceptable. Without cause and mechanism a phenomenon is by definition **supernatural magic**. It leaves what could hopefully be a useful effect essentially unreliable. It also makes modifying and improving, engineering it more difficult.

The purpose of this work is to investigate of why and how entanglement operates. The effects have the appearance of validly existing therefore, there must be an operative mechanism for it.

RESOLUTION OF THE PROBLEM OF REALISM

Particles in Quantum Mechanics

Quantum mechanics postulates that the *state* of every elementary particle can be described by a *wave function*, a mathematical representation from which one can calculate probabilities that the particle is to be found in a particular location or state of motion; and that the act of *measurement / observation* of the particle causes the calculated set of probabilities to *collapse* to the value found by the measurement.

70

In <u>Quantum Mechanics</u> the condition that, until *measurement / observation*, the specific *state* of the particle is undetermined, consisting of various probabilities of various states according to the *wave function*, is also described as that the particle is in a *superposition* of all of the states.

<u>Realism</u> is the principle that all objects must objectively have a pre-existing value of any of their measureable characteristics independent of any measurement that is made and before the measurement is made. The measurement (observation) cannot and does not create or initiate the value. This means that every material object exists independent of being observed.

But, based upon results like experiment #5 of the Double Slit Experiment, below, Quantum Mechanics' position is that a particle is merely a probability wave function having no material existence until it is "observed". Observation of the particle then causes the wave function to "collapse" to one of the probabilistically superposed states.

What that is telling us is that nothing is real until it has been observed; that we cannot say anything about what things are doing when we are not looking at them; that they do not exist to do anything when we are not looking at them.

This has caused some very well respected cosmologists (*e.g.* Stephen Hawking) concern that this implies that there must actually be something 'outside' the universe to look at the universe as a whole and collapse its overall wave function for the universe to be.

It is not reasonable and is another violation of the principle of Occam's Razor [that the simplest explanation is the most likely] that we should need two different laws to explain the behavior of objects depending on how large or small they are. Why should the laws of cause and effect of the macro world not apply to the micro ?

The following analysis of the Double Slit Experiment resolves these problems. The form of light used in the experiment is important and is a significant factor in the interpretation of results. Consequently the experiment must be analyzed twice: once with the incident light being "natural light" as was certainly the case before the invention of the laser in 1960, and once with the incident light being "coherent light" as in contemporary modern experiments. The "coherent light" case applies for both the incident light being light or particles such as electrons or protons.

The analysis, beginning on the next page, compares the Quantum Mechanics interpretation of each phenomenon with the classical interpretation. It shows that all of the Quantum phenomena are fully explained in terms of classical physics with no need nor involvement of interpretations like the Copenhagen and no "weird behavior" that lacks substantiation by applicable mechanism.

THE DOUBLE SLIT EXPERIMENT - [AS BEFORE 1960]

The incident light is "natural light". Deeming it to be a particle leads to the "spookiness" and contrary-to- common-sense behaviors. Seeing the photon as the wave it is resolves all of those problems.

Images for the light behaving as waves:

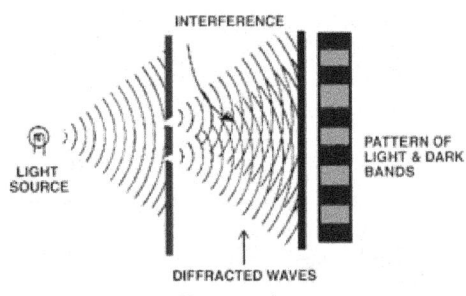

Figure A
Experimental Set-Up and Both Slits Open

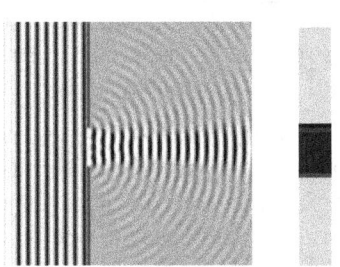

Figure B
Same Set-Up, Only One Single Slit Open

Images for the light behaving as particles:

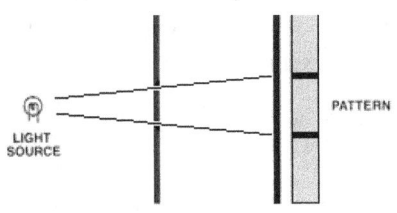

Figure C
Experimental Set-Up and Both Slits Open

Figure D
Same Set-Up, Only One Single Slit Open

Nature of the Propagations	Quantum Mechanics - Takes Light to be Particulate Photons	Classical Physics Light is "Natural Photons"
The light source is "natural light"	Photons are particle in nature. They do not extend over a broad wave front but rather are analogous to bullets, able to freely pass through narrow slits, and travel in one specific direction. Figure C or D. Their behavior is wave or particle depending on circumstances.	Light is purely wave in nature. The light waves are of broad wave fronts and spread out in space as shown in Figure A. They are not spatially limited to small narrow particles. Encountering a slit the part of the broad wave front at the slit passes through, the rest does not.

The Experiments	Comparison Quantum Mechanics vs. Classical Physics	

As Done
[#1] Light: Continuous
      Slits: Both open
Actually Happens
   Wave interference pattern Figure A
	Quantum Mechanics Interpretation
	
	[#1] Particulate photons pass through both slits. Their wave behaviors there diffract and interfere making a Figure A effect.
	
	BELOW "WEIRD BEHAVIOR" IS IN ITALICS
		Classical Physics Interpretation
		
		[#1] Part of the broad wave front enters both slits, there diffracts and the results interfere Figure A.

As Done	Quantum Mechanics Photons	Classical Physics Interpretation

[#2] Light: Repeated single photons
    Slits: Both open
 	
[The photons are forced to the one-at-a-time mode by reduction of the light intensity.]
 	
Actually Happens
   "Implied" wave interference pattern
	[#2] We would expect a single photon to go through one slit or the other Figure C; it cannot go through both at the same time to create an interference pattern. Rather the expectation is a Figure C spot opposite each slit. But what we get is an "implied" interference pattern as Figure A.
	
	It is as if each individual photon "knows" that both slits are open even though it passes through only one.
	
	And it will place itself on the screen such that when enough have passed through they have built up an implied interference pattern, in spite of that there cannot possibly be any actual interference of the one-at-a-time photons.
		[#2] The single photon aspect is the same broad wave front with the waves in single bursts of $W = h \cdot f$.
		
		The light waves are at broad wave fronts not spatially limited to small particles like the photons. The broad wave front passes both slits, and the results interfere as Figure A.

As Done	Quantum Mechanics Photons	Classical Physics Interpretation

[#3a] Light: Repeated single photons
    Slits: One open, one closed
Actually Happens
   The photons cluster around points on the detector screen behind the open slit Figure D there being no 2nd wave to interfere with.
 	
[3b] But, if the second slit is opened
 	
Actually Happens
   The photons immediately start to form an "implied" interference pattern as in [#2] Figure A.
	[#3a] We expect a single photon to go through one slit or none. The expectation is a Figure D spot opposite the open slit and that is what happens.
	
	[#3b] *But, an individual photon passing through one of the slits is not only aware of the other slit, but also is aware of whether or not it is open.*
	
	If the closed other slit is opened it immediately reverts to the [#2] situation Figure A.
	
	It is as if each individual photon "knows" that both slits are open or that just one slit is open even though it passes through only one and doesn't visit the other to test it, and it will place itself on the screen in such a position that when enough have passed through they have collectively built up an interference pattern, in spite of that there cannot possibly be any actual interference of the one-at-a-time photons.
		[#3a] The light waves are at broad wave fronts not spatially limited to small particles like the photons. They enter the open slit and produce a spot as in Figure B.
		
		[#3b] If the closed slit is opened the broad wave front enters both slits, there diffracts and the results interfere.
		
		The one-at-a-time aspect is the same broad wave front with the waves in single bursts of $W = h \cdot f$.

73

As Done	Quantum Mechanics Photons	Classical Physics Interpretation
[#4a] Light: Repeated single photons Slits: Both open Detectors at both slits [Detectors observe which slit a photon goes through but let it pass on to the screen.] *Actually Happens* The pattern of Figure C, *a spot opposite each slit with no interference pattern.* [#4b] Remove the detectors *Actually Happens* Reverts to [#3b] The photons immediately start to form an "implied" interference pattern as in [#2] Figure A.	[#4a] Because of the one-at-a-time single photons a photon can only go through one slit or the other. There will never be a pair of photons at the same time. Thus there can never be "real" interference. *Being "observed" the photons are forced to collapse to particulate form unable to form an implied interference because of not being waves but particles..* [#4b] We expect a single photon to go through one slit or the other Figure C; it cannot go through both at the same time to create an interference pattern. The expectation is a Figure C spot opposite each slit. But what we get is an "implied" interference pattern as Figure A. *It is as if each individual photon "knows" that both slits are open even though it passes through only one. And it will place itself on the screen in such a position that when enough have passed through they have collectively built up an implied interference pattern, in spite of that there cannot possibly be any actual interference of the one-at-a-time photons.* [#4a] & [#4b] *The photons "know" whether the detectors are present or not. They adjust their behavior accordingly.*	[#4a] The one-photon-at-a-time light propagation has the same broad wave front as Figure A not the narrowness of a "bullet like" photon. The reduced intensity wave front (reduced to produce single photons) arrives in single bursts of energy $W = h \cdot f$ with periods of non-wave separating successive bursts. Each burst is a half cycle of an electromagnetic sinusoid as in "Analysis of the Photon from Its Generating Source" in Section 2. #4a] & [#4b] In detecting a part of a wave burst passing through its slit the detector there unavoidably slightly delays that part of the burst's wave front.(see * below). The result is: [a] A reported detection and [b] The delayed part of the burst proceeds from its detector to the screen alone without interference. [c] The absence of interference is because the portion of the burst's overall wave front that passes the other slit does so at a different time than at the first slit because of the slight time difference in the burst's delay at the other slit's detector. (see * below). With that action because of the non-identical detectors a wave passes on from its slit at a time that that is not happening at the other slit making each slit the case of Figure B, the wave version of Figure C. It is the process of detection and its time delay variation that produces the change from the interference patterns to "spots". [#4b] With the detectors removed the two slits wave bursts are again synchronized and interfere.

As Done	Quantum Mechanics Photons	Classical Physics Interpretation
As Done [#5a] Light: Repeated single 　　photons 　Slits: Both open 　Detector at just one of the two 　　slits *Actually Happens* 　　The pattern of Figure C, *a spot* 　　opposite each slit with no 　　interference pattern. *As Done* [#5b] Light: Repeated single photons 　　Slits: Both open 　　No detectors *Actually Happens* 　An "implied" interference pattern 　　　　　　　as in Figure A Note: Experiment #5 is identical to #4 except in #4 there are detectors at both slits and in #5 a detector at just one slit.	[#5a] & [#5b] *If a photon passes through a slit that does not have a detector, it not only "knows" if the other slit is open or not, it "knows" if the other slit is being observed.* 　*If there is no detector at the other slit as well as the one it is passing through, it will produce a Figure A interference pattern.* 　*Otherwise it will act as a Figure C particle.* *This is a specific example of the interaction of the observer with the experiment. If we try to observe the action, it collapses into a definite particle, but when we do not observe it opts between wave and particle depending on the situation at the other slit, which it "knows".* *The "collapse of the wave function" theory seeks to explain how an entity such as a photon or an electron, could 'travel as a wave but arrive as a particle'. The theory proposes that what passes through the slits is not a material wave nor particle but a "probability wave".* *In this theory, an electron or photon that is not being observed does not exist as a particle at all, but has a wave-like property covering the areas of probability where it could be found.* *Once the electron or photon is observed, the wave function is forced to collapse because the various probabilities have become something definite; the electron or photon becomes a particle.*	[#5a] & [#5b The same as for #4, above. The effect of just one detector at one slit produces the timing separation delays that prevent the flow through both slits jointly interfering in the same fashion as with detectors at both slits. *A detector cannot avoid causing modification of that which it detects, delaying or changing it in some manner. Furthermore, no two detectors will do so identically. The difference need merely be the wave burst's half-period. 　The half period of visible light is in the range of 10^{-5} m. At the speed of light, $3 \cdot 10^{8}$ m/s the delay needed to separate the bursts at the two slits is a delay on the order of 10^{-14} s. Much greater delays and variations in that amount are to be expected from the detectors

THE DOUBLE SLIT EXPERIMENT - [AS AFTER 1960]

The incident light is laser-generated "coherent light".

Images for the light behaving as waves:

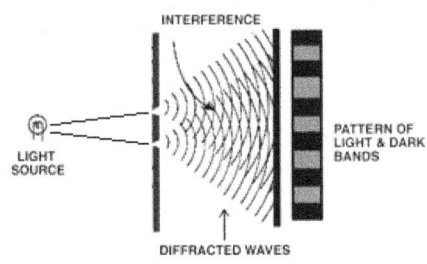

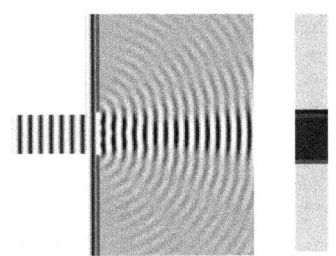

Figure A
Experimental Set-Up and Both Slits Open

Figure B
Same Set-Up, Only One Single Slit Open

Images for the light behaving as particles:

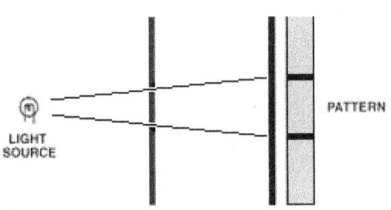

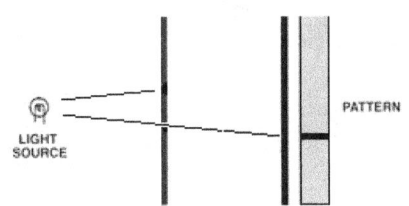

Figure C
Experimental Set-Up and Both Slits Open

Figure D
Same Set-Up, Only One Single Slit Open

Nature of the Propagations The light source is "coherent light"	Quantum Mechanics - Takes Light to be Particulate Photons Photons are particle in nature. They do not extend over a broad wave front but rather are analogous to bullets, able to freely pass through narrow slits, and travel in one specific direction. Figure C or D. Photons exhibit a combination of wave behavior or particle behavior depending on the circumstances.	Classical Physics - Light is Collimated "Natural Photons" Light is a stream of half-cycle sinusoidal bursts of energy $W = h \cdot f$. The initially multi-directional bursts are focused into a narrow Collimated stream as if a particle.

The Experiments	Comparison Quantum Mechanics vs. Classical Physics	

As Done
 [#1] Light: Continuous single photons
       Slits: Both open
 Actually Happens
   Wave interference pattern Figure A
	Quantum Mechanics Interpretation
	   Particulate photons pass through both slits. Their wave behaviors diffract and interfere as shown [Figure A].
	
	 BELOW "WEIRD BEHAVIOR"
	 IS IN ITALICS
		Classical Physics Interpretation
		   Half-cycle sinusoidal bursts pass through both slits. Their waves diffract and interfere as shown [Figure A]

	Quantum Mechanics Photons	Classical Physics Interpretation
As Done [#2] Light: Repeated single photons Slits: Both open *Actually Happens* "Implied" wave interference pattern	We would expect a single photon to go through one slit or the other Figure C; it cannot go through both at the same time to create an interference pattern. Rather the expectation is a Figure C spot opposite each slit. But what we get is an "implied" interference pattern as Figure A. *It is as if each individual photon "knows" that both slits are open even though it passes through only one.* *And it will place itself on the screen such that when enough have passed through they have built up an implied interference pattern, in spite of that there cannot possibly be any actual interference of the one-at-a-time photons.*	The half-cycle sinusoidal bursts are diffracted at their slits. Again just as with the waves of the "natural light" Figure A, the diffracting spreads out the wave front destroying the coherence. Depending on where on the slit encounters where on the half-cycle sinusoid the new directions of the diffracted sinusoids vary. Successive bursts achieve simultaneity by the sideward diffracted wave from one slit arriving at the other slit when that other is passing a burst. Successive various diffracted sinusoids mark out an interference pattern on the screen.
As Done [#3a] Light: Repeated single photons Slits: One open, one closed *Actually Happens* The photons cluster around points on the detector screen behind the open slit Figure D. *As Done* [3b] But, if the second slit is opened *Actually Happens* The photons immediately start to form an "implied" interference pattern as in [#2] Figure A.	[#3a] We expect a single photon to go through one slit or none. The expectation is a Figure D spot opposite the open slit and that is what happens. [#3b] *But, an individual photon passing through one of the slits is not only aware of the other slit, but is aware of whether or not it is open.* *If the closed other slit is opened it immediately reverts to the [#2] situation Figure A.* *It is as if each individual photon "knows" that both slits are open or that just one slit is open even though it passes through only one and doesn't visit the other to test it, and it will place itself on the screen in such a position that when enough have passed through they have collectively built up an interference pattern, in spite of that there cannot possibly be any actual interference of the one-at-a-time photons.*	[#3a] Each half-cycle sinusoid behaves at the open slit analogously to Figure B. [#3b] Reverts to [#2]

As Done	Quantum Mechanics Photons	Classical Physics Interpretation
[#4a] Light: Repeated single photons Slits: Both open Detectors at both slits [Detectors observe which slit a photon goes through but let it pass on to the screen. *Actually Happens* The pattern of Figure C, *a spot opposite each slit with no interference pattern.* *As Done* [#4b] Remove the detectors *Actually Happens* Reverts to [#3b] The photons immediately start to form an "implied" interference pattern as in [#2] Figure A.	[#4a] Because of the one-at-a-time single photons a photon can only go through one slit or the other. There will never be a pair of photons at the same time. Thus there can never be "real" interference. *Being "observed" the photons are forced to collapse to particulate form unable to form an implied interference because of not being waves but particles..* [#4b] We expect a single photon to go through one slit or the other Figure C; it cannot go through both at the same time to create an interference pattern. The expectation is a Figure C spot opposite each slit. But what we get is an "implied" interference pattern as Figure A. *It is as if each individual photon "knows" that both slits are open even though it passes through only one. And it will place itself on the screen in such a position that when enough have passed through they have collectively built up an implied interference pattern, in spite of that there cannot possibly be any actual interference of the one-at-a-time photons.* [#4a] & [#4b] *The photons "know" whether the detectors are present or not. They adjust their behavior accordingly.*	#4a] & [#4b] In detecting a wave burst passing through its slit the detector there unavoidably slightly delays it.(see * below). The result is: [a] A reported detection and [b] The delayed burst proceeds from its detector to the screen alone without interference. [c] The absence of interference is because the burst that passes the other slit does so at a different time than at the first slit because of the slight time difference in the burst's delay at the other slit's detector. (see * below). With that action because of the non-identical detectors a wave passes on from its slit at a time that that is not happening at the other slit making each slit the case of Figure B, the wave version of Figure C. It is the process of detection and its time delay variation that produces the change from the interference patterns to "spots". [#4b] With the detectors removed the two slits wave bursts are again synchronized and interfere.

As Done	Quantum Mechanics Photons	Classical Physics Interpretation
[#5a] Light: Repeated single photons Slits: Both open Detector at just one of the two slits *Actually Happens* The pattern of Figure C, *a spot opposite each slit with no interference pattern.* *As Done* [#5b] Light: Repeated single photons Slits: Both open No detectors *Actually Happens* An "implied" interference pattern as in Figure A Note: Experiment #5 is identical to #4 expect in #4 there are detectors at both slits and in #5 a detector at just one slit.	*[#5a] & [#5b] If a photon passes through a slit that does not have a detector, it not only "knows" if the other slit is open or not, it "knows" if the other slit is being observed.* *If there is no detector at the other slit as well as the one it is passing through, it will produce a Figure A interference pattern.* *Otherwise it will act as a Figure C particle.* *This is a specific example of the interaction of the observer with the experiment. If we try to observe the action, it collapses into a definite particle, but when we do not observe it opts between wave and particle depending on the situation at the other slit, which it "knows".* *The "collapse of the wave function" theory seeks to explain how an entity such as a photon or an electron, could 'travel as a wave but arrive as a particle'. The theory proposes that what passes through the slits is not a material wave nor particle but a "probability wave".* *In this theory, an electron or photon that is not being observed does not exist as a particle at all, but has a wave-like property covering the areas of probability where it could be found.* *Once the electron or photon is observed, the wave function is forced to collapse because the various probabilities have become something definite; the electron or photon becomes a particle.*	[#5a] & [#5b] The same as for #4, above. The effect of just one detector at one slit produces the timing separation delays that prevent the flow through both slits jointly interfering in the same fashion as with detectors at both slits. *A detector cannot avoid causing modification of that which it detects, delaying or changing it in some manner. Furthermore, no two detectors will do so identically. The difference need merely be the wave burst's half-period. The half period of visible light is in the range of 10^{-5} m. At the speed of light, $3 \cdot 10^8$ m/s the delay needed to separate the bursts at the two slits is a delay on the order of 10^{-14} s. Much greater delays and variations in that amount are to be expected from the detectors

79

Examining in the above tables the right two columns, "Quantum Mechanics Photons" [QM] and "Classic Physics Interpretation" [CP], the QM is quite complex and the CP is simple. In the QM column the comments in italics are all various instances of unsound, unscientific, un-validated Quantum "weirdness" contentions. In every case the CP column provides a sound classical scientific analysis of the behavior.

Of note is that CP uses none of and QM depends on use of unscientific and un-validated descriptions referring to a photon being in some sense aware of states and events beyond its ken, it having consequent options and the ability to choose between them, and doing such things while not really existing until it subsequently is 'collapsed' by being 'observed'.

In particular the QM description and analysis of what is going on requires the particle to be in the form of a probability wave function until its being "observed" collapses it to a specific state. That the CP explains the behavior without any of that speculative and unsupported-by-evidence Copenhagen interpretation demonstrates the non-validity of the Copenhagen interpretation and similar other interpretations.

THE DOUBLE SLIT EXPERIMENT WITH MATTER PARTICLES INSTEAD OF PHOTONS

In Section 8 it was shown that every matter particle, that is every Spherical-Center-of-Oscillation, emits oscillatory Propagated Outward Flow radially in all directions and when in motion is accompanied by an oscillatory matter wave so that for the purposes of these experiments those matter particles are equivalent to coherent photons. All of the Double Slit phenomena presented derive from and are applicable to the Quantum behavior of matter particles.

CONCLUSION OF THE PROBLEM OF REALISM

1 – In brief, CP makes sense and QM seems like ***supernatural magic*** in: supplying no cause nor mechanism for its contended superposition of states, nor for its particles existing only as a probability wave function, nor for the particles' change from probability to specificr existence initiated by an act of observation in some form, and nor for the particles having unexplainable knowledge about events and conditions physically beyond their ken.

2 – The above analysis of the Double Slit Experiment demonstrates that particles that can exist do so whether observed, measured or not. They exist not as wave functions having material reality only when collapsed by being observed but as not-a-problem <u>Realism</u>'s *Spherical-Centers-of-Oscillation* propagating their *outward flow.*

3 – The Copenhagen and similar interpretation of particles being only a wave function before existence upon being observed fails as:

> [a] not necessary to explain or account for phenomena, as shown above by CP, and

> [b] unscientific and unsound in that it lacks any cause or mechanism for the behavior.

4 – Wherefore: the classical principle of <u>Realism</u> is sustained.

Particle Centers-of-Oscillation and Quantum Mechanics

Because *Spherical Centers-of-Oscillation* oscillate over a cyclic range of instantaneous values per the particular *[1 – Cosine]* waveform of each case, the *state* of the particle can be thought to so continuously vary cyclically. That would be analogous to the Quantum Mechanics "*state*" of the particle as the particular instantaneous position in its wave function as analogous to the particular instantaneous position that its *Spherical Centers-of-Oscillation* is at a particular moment.

- The waveform of the *Spherical Centers-of-Oscillation* would be the "*wave function*" of Quantum Mechanics.

- The *Spherical Centers-of-Oscillation's* oscillation over its range of instantaneous values would be the Quantum Mechanics described behavior that particles are in a *superposition of all possible states* until a *measurement / observation* causes the *superposition* to c*ollapse* to the state *measured / observed.*

- Of course the particle's oscillation is in only one pure, single, simple, state, point in its cyclic oscillation at any moment not simultaneously in all possible states as the Quantum Mechanics contention implies;

- The *collapse* would be the selection of that particular instantaneous position of the waveform of the *Spherical Centers-of-Oscillation* that it happens to occupy at the instant of the *measurement / observation*.

■ - <u>However</u> none of that means anything like the imaginative Quantum Mechanics' non-existence of the particle until it is actually observed.

REALITY AND REALISM

Developments in the progress of physics must always pass two tests.

<u>The first test</u> is the test of standing up to examination in terms of the real world in which our reality exists. That requirement applies equally to the evaluation and interpretation of material experiments, of "thought experiments" and to the related mathematics.

For example:

The electrons flowing in copper wires are atomic level particles that according to Quantum Mechanics' denial of <u>Realism</u> should be represented only by their wave functions until they are observed [measured]. That means that all electrical activity must be null until its collapse into existence. Every day experience shows that that is not the case.

It may be contended that all electrical activity is in fact always observed or measured by producing the effect or action that it was designed to do and is operated to do. But, can that be true of the device pilot light that is on all night when no one knows, sees, or cares?

And, if "yes the wave function and collapse principle includes even that", then while the principle can be deemed valid if one so wishes it would be a *reductio ad absurdum* in that everything is now and always observed and therefore collapsed.

Quantum Mechanics' denial of Realism fails the test of real world reality.

<u>The second test</u> is mechanism. For particles to be solely a "superposition of all of the possible states of the particle" and then a "collapse to the specific state measured" there must be a cause of both the superposition and of the reversion to one particular specific state, there must be a mechanism to account for each.

Again, Quantum Mechanics denial of <u>Realism</u> fails the test of supporting cause and mechanism.

Resolving the Problems of Locality and Entanglement

THE PROBLEM

Locality states that an object is only directly influenced by its immediate surroundings. For an action at one location to have an influence at another non-contiguous location, something in the space between the locations must mediate the spatial separation.

Entanglement is a joint state of two or more particles where one particle instantly "knows" what happens to and what is the state of the other and appears to be able to force a change in the state of the other, even though there appears to be no means for such communication between the particles, which may be separated by arbitrarily large distances.

Clearly the two are opposed. One claims that the spatially separate have no interaction unless something intervenes to mediate the separation. The other claims that the spatially separate can significantly interact without any intervening mediation. Either one or the other might be valid but both cannot be simultaneously valid. Which is it?

Quantum Mechanics cites the EPR Paradox type experiments as proof of Entanglement.

THE EPR PARADOX.

The EPR Paradox (or Experiment) is so named because it was a thought experiment devised by Einstein, Boris Podolsky and Nathan Rosen in 1934-1935. In 1976 and subsequently the experiment has been physically run. The results have always been interpreted as favoring 'non-locality', the opposite of classical 'locality'.

In the experiment a pair of protons, for example, associated with one another in a singlet state will always have a total angular momentum of zero, as they each have equal and opposite amounts of spin.

According to Quantum Mechanics, for each of the protons its probability wave will not collapse and its specific spin be decided until it has been measured (observed). If one measures the spin of one proton, according to quantum theory, the other proton instantly "knows" and adopts the opposite spin its probability wave having simultaneously likewise collapsed.

Separating the particles in opposite directions and measuring one of them for spin has been carried out over a distance of 10 km. The instant it is measured and the spin determined, the other particle apparently adopts the opposite spin. The time interval is zero, instantaneous.

It would appear that something is communicating between the particles and at light speed or faster. But, what and how ? We must identify and understand the mechanism producing the observed quantum entanglement effects or else find and demonstrate that the effects are not real and only apparent. Otherwise, failing those two, the observed quantum entanglement effects are real as observed and interpreted but have no cause, no mechanism and, therefore, are, by definition, *supernatural magic* which is scientifically unacceptable.

ANALYSIS OF ENTANGLED QUANTUM EFFECTS COMMUNICATION

Entanglement is defined by the following: If two particles are in a state such that there is a matching correlation between two canonically conjugate dynamical quantities, they are termed as being "entangled". Such entangled behavior has been noted in instances, for example, of particle angular momentum and of photon polarization.

The correlation means that there is "coherence" among the entangled particles. When the coherence is lost the particles are "decoherred".

According to Quantum Mechanics any measurement of a property of a particle causes an irreversible collapse of its wave function to the just measured quantum state of the particle. In the case of entangled particles, the effect of such a measurement will be on the entangled system as a whole.

Requirements for Entanglement Mechanism

For the entangled "matching" to be maintained there must be a communication among the entangled particles, there must be something flowing from each to all of the others so that each has the necessary information to determine what its specific matched correlated state must be.

Because the entangled particles are continuously in motion, curvilinear or oscillatory about a location, directing the communicating flow from particle to particle is impractical because the location to which to send the outgoing communication is indeterminate. Therefore, it cannot be directed to solely those of the entanglement. The only alternative is that the communication must be generally broadcast.

Because any particle might be called on to participate in an entanglement, every particle, all particles, must be continuously broadcasting their quantum state so that a means is required to leave non-entangled particles unaffected by the operating of the entanglement matching action..

In order for the communicated enforcement to affect only the entangled particles, when two or more particles are entangled there must be some kind of entanglement identification mark or notation on each entangled particle, the mark identifying each as part of a system of entangled particles and in that role in its part in the "matching correlation between the conjugate dynamical quantities" of its entanglement being maintained.

Lacking that identifying mark a particle is not involved in a "matching correlation" and is not entangled.

Furthermore there must be something enforcing the maintenance of the correlation at each entangled particle. For example in the case of entangled particles "A" and "B", upon a change in the state of "A" a communication must be sent to "B" and: either [a] that communication itself from "A" received at "B" has a mechanism to cause or force "B" to change its state to the new correlation, or [b] the entangled "B" itself has such a mechanism to cause or

84

force itself to change its state to the new correlation that "B" mechanism being triggered into action by the received communication from "A ".

Either that mechanism is inherent in every particle or else it must be placed into each particle upon initiation of the entanglement and removed upon each decoherence.

In summary entanglement involves the following requirements:

1 - Every particle must be broadcasting information as to its current quantum state;

2 - Each entangled particle must have an identification to that effect;

3 - For each entangled pair there must be mechanism that enforces "matching correlation".

There is, in fact a communicating flow from every particle to every other particle in the universe. That flow, developed in the earlier sections 6 and 7 is described as follows.

The Flow from Particle Centers-of-Oscillation

In Section 6 it is found that there is a spherically outward flow of oscillatory *Medium* wave from every particle. That existing flow is the only possible candidate for the communicating flow to be the mechanism for entanglement because it already exists. [That *Propagated Outward Flow* has a primary role in the mass of particles and in the action of Coulomb's Law and Ampere's Law, see Reference [2]. A second universally broadcast such flow in addition is not possible because it would interfere with the existing *Propagated Outward Flow.*]

That Which is Flowing

- Contemporary particles are Big Bang successors of the original *[1 - Cosine]* oscillations with which the universe began. Thus the outward flow of the original oscillations is a property of present particles. That which is flowing is the same original primal *Medium*, the substance of the original oscillations, as at the beginning of the universe.

- Since it is flowing outward from each of the myriad particles of the universe simultaneously and since that flow is interacting with the myriad other flows of those particles without untoward interference, the *Medium* must be extremely intangible for all of that to take place. Any one particle's flow flowing largely freely through that of other particles, is as intangible as … well , "field".

The Oscillatory Medium Flow

- The initial medium supply of the universe, oscillating in *[1 - Cosine]* form, came into existence at the Big Bang. Therefore the initial medium supply of each particle, each being a direct "descendant" of the original oscillation at the universe's beginning, must be likewise oscillatory in form. Therefore the radially outward flow from each particle is likewise an oscillatory medium flow of the same *[1 - Cosine]* form.

- For such a flow to persist there must be a supply of that outward flowing substance in every particle. And, for that flow to have persisted the billions of years since the "Big Bang" that "supply" must be an extremely concentrated reservoir of that which flows outward [concentrated relative to the outward flow].

- The reservoir is the spherical "core" of radius δ, equation *(6-10)* at the center of all particles.

In Section 7 it is found that the flow from particles in motion is forced to be different in the forward, rearward, and sideward directions, see Reference [2]. Those differences overall carry information about the state of the particle including its direction, velocity, energy, frequency and mass. But particles encountering that flow experience only a very small portion of the total spherical wave front propagated by the particles, a sample that may be of its forward only, or sideward only, or rearward only, or whatever part of the total picture. Without reference to the overall wave front the minor sample in an actual encounter carries negligible information of use for the quantum entanglement problem.

Thus the actual communicating flow satisfies the quantum requirement #1 of broadcast communication but fails to satisfy requirements #2 and #3, entanglement identification and means to enforce the correlation of quantum states. Therefore, there is no flow satisfying the communication requirements for entanglement or there must be a second universal broadcast flow from all particles fulfilling the requirements.

However, any second universal flow would, by the presence of its own "core" interfere with the onmi-directional role of the earlier original "core". Consequently, there is no facility to perform the communication requirements for entanglement.

Conclusion of Analysis of Quantum Effects Communication

Having now found that there is no mechanism able to support or justify the Quantum Mechanics interpretation of entanglement effects it remains to investigate whether those effects are real or only apparent because science cannot accept their being real but lacking cause and mechanism, which is the definition of ***supernatural magic***.

ANALYSIS THAT THE ENTANGLED QUANTUM EFFECTS ARE NOT REAL

In terms of Quantum Mechanics, for example in the EPR paradox, before the measurement and consequent "collapse" where are the protons' angular momentums [spin angular momentums] located, or how are they expressed ? The only place available is in each proton's wave function because that is the only thing specifically related to each proton. The two protons don't specifically "exist" until their probability waves collapse. But, the wave function is non-material and cannot contain angular momentum, only probability. If the two particles' angular momentums do not exist somewhere before the measurement and its "collapse" they cannot exist afterward. The "collapse" cannot call them into existence from nothing.

Quantum Mechanics would argue that before the "collapse" the protons do exist "in a superposition of all possible states for them". But then, where is their angular momentum ?

- Is it in full in each of those superposed states ready for any one of them to be selected at the "collapse" ? Then how did only one protons worth of angular momentum get so enlarged ? And what of the left over angular momenta in the multitude of states not selected at the "collapse" ?

- Is it spread out allocated among the various possible superposed states ? How can, how does that happen ?

- Is it only in the state to which the superposition collapses ? But which state that is is not determined until the instant of collapse ...

... and, if that state were earlier determined what point would there then be to the variety of other states in the superposition, states then with no function nor possibility of reality ?

86

Therefore the two protons must, and do, "pre-exist" with opposite spins in correlation as if already "collapsed" **before the first measurement**. Thus there is no probability wave to "collapse" because the state that results from collapse is already extant.

[It has already in the prior section been found that Realism is valid, particles exist independent of whether observed or not, and there is no "collapse of the wave function" to bring particles into existence.]

The problem of instantaneous communication goes away with the protons in "existence" and spinning oppositely in their correlated spins **before** the first measurement. Then measuring one and finding the other instantaneously correlated is assured because it was "pre-assured".

There is no question nor issue of "action at a distance" at all here because there is no actual "action". Being a contended example of entanglement, this experiment actually supports Locality and denies Entanglement. The above critical analysis applies as presented for protons, electrons, photons and all such atomic scale particles. [In the case of photons the entangled property is angle of polarization, not angular momentum.]

SUMMARY OF ANALYSIS

The previously above Section 9 conclusion of the problem of Realism has found that Realism is valid, particles do exist independently of whether observed or not, and there is no "collapse of the wave function" to bring particles into existence. That means that the wave function is only a mathematical invention and does not exist in material reality.

In the EPR experiment before any first particle measurement the entangled particles all "existed" and were each in its proper state correlated with the others. They were material particles not "wave functions". They did not need "collapse" to come into material existence and they would experience no "collapse" because there was no probability wave there to collapse. They all are in a set of correlated states after the measurements as before.

Consequently their separation distances were of no concern and there was no "spooky" action, no violation of Locality..

PART V -- ANALYSIS OF "SPIN" AND QUANTIZED ANGULAR MOMENTUM

In discussions of Quantum Mechanics a property of particles identified as "spin" and involving angular momentum occurs frequently as for example a referring to "spin" up or "spin" down as quantum angular momentum 'states'. In those discussions it is often stated that no specific rotary motion (spin) is necessarily involved but that rather some intrinsic property of the particle being treated, an electron or an atom, is what is intended.

The intent is that quantized angular momentum is a natural property of particles such as electrons or atoms and that is contended in spite of there being no cause or mechanism for the particles to have that property and with the actual denial that any physical spin as rotation about a central axis is present.

That contention is defended by the citing of three different experimentally revealed behaviors:

      · atomic spectra fine and hyperfine structure;

      · atomic electron specific stable versus unstable orbits;

      · the Stern-Gerlach experiment.

That there is no quantized angular momentum "natural property of particles" quantum or otherwise and that the three behaviors listed just above are completely explained by classical mechanics proceeds as follows.

      - Atomic Spectra Fine and Hyperfine Structure

      -Atomic electron stable orbits

      - The Stern-Gerlach Experiment

            and

      - The End of Particle 'Spin' and Its Quantized Angular Momentum

Atomic Spectra Fine and Hyperfine Structure

THE PROBLEM

In discussions of Quantum Mechanics a property of particles identified as "spin" and involving angular momentum occurs frequently. In those discussions it is often stated that no specific rotary motion (spin) is necessarily involved but that rather some intrinsic property of the particle being treated, an electron or an atomic particle, is what is intended. Yet the calling upon angular momentum quantized into two alternative states ('spin' up or 'spin' down) is part of those presentations.

The Stern–Gerlach experiment [see Section 13, below] appeared to demonstrate that the spatial orientation of angular momentum is quantized, and thus was an atomic-scale system having intrinsically quantum properties. This experiment was decisive in (wrongly) convincing physicists of the reality of angular-momentum quantization in all atomic-scale systems as for example the stable electron orbits in atoms [see Section 12].

That immediately raises the questions: What is it that makes the difference. (e.g. between 'spin' up versus 'spin' down).? Why are there just the two states not more ? What is going on with this quantization of angular momentum. ? What is its mechanism ?

The purpose of this section and the following two is to demonstrate that quantized angular momentum is not a "natural property of particles" and that the three behaviors contending that it is

· *atomic spectra fine and hyperfine structure;*
·*atomic electron specific stable versus unstable orbits;*
·*the Stern-Gerlach experiment.*

are completely explained by classical mechanics.

FINE STRUCTURE AND SPIN

When the line spectrum of Hydrogen is obtained with a spectrometer of high resolving power it is found that the lines that appear as simple single lines at low resolving power are in fact pairs of lines. This phenomenon is referred to as the *fine structure*. The splitting of the (low resolution) single line into (high resolution) two lines is on the order of about 1 part in 10^4. Sommerfeld addressed this problem showing that if the orbital electrons had elliptical orbits, in which the electron velocity would be relatively slow far from the nucleus and faster than for the circular orbit case near the nucleus, the relativistic mass increase at the higher velocity provided a minute energy increase that was on the order of the correct amount to account for the line splitting. That is, the elliptical orbit's energy would be slightly greater than a circular orbit's energy.

91

Sommerfeld's model for how the fine structure arises, a model based upon the conceived direct motion and action of the electrons, was soon superseded by Quantum Mechanics, a model that sought not to directly represent electron motion but rather to express the electron behavior and its effects. However, in spite of the wide spread acceptance of Quantum Mechanics, the concept of elliptical electron orbits has been retained.

Quantum Mechanics overthrew the Bohr-Sommerfeld theory shortly after its development. In Quantum Mechanics the fine structure is attributed to the interaction of the magnetic field due to the electron's spin on its own axis with the magnetic field due to the electron's orbit around the nucleus. This is referred to as spin-orbit coupling. The two cases that are contended to account for the two lines close together in the Hydrogen spectrum are for the electron's spin angular momentum vector in the same direction as the orbital motion angular momentum vector or in the opposite direction.

In a sense the conception that traditional 20th Century physics had of the electron is of a powder of negatively charged minute specks compressed into a little ball. (One of their concerns was that of what holds the electron together; with all of that charge packed so closely why does it not explode ?) In that sense, the electron is conceived of as spinning on its axis. It is conceived that the consequent circular motion of the specks of charge that are rotating about the electron's spin axis constitute a small current and generate a small magnetic field.

Actually, traditional 20th Century physics did not know, and had no way of knowing, whether the electron spins or not and if so then how rapidly, how (in traditional 20th Century physic's terms) the charge is distributed throughout the electron and what the electron diameter is, and so forth, all data necessary to calculation of its spin magnetic field. The contention of electron spin and its associated magnetic field depends entirely on that the concept is used to explain a fine characteristic in atomic line spectra. The amount of spin and the amount of consequent magnetic field was taken to be that at the value that explained the spectral fine structure.

But, for *Spherical-Centers-of-Oscillation* there can be no such concept. A c*enter-of-oscillation* cannot spin because of the nature of its structure and function.

Fine structure is the result of each orbital electron's having one or the other of two possible slightly different energy states in its orbit. In traditional 20th Century physics the two energy states result from the electron spin angular momentum (and magnetic) vector being in the same or opposite direction relative to the orbital motion angular momentum (and magnetic) vector. Spin in fact not being the cause because there is no spin, there must be some other cause that produces the same effect.

THE EFFECT OF ABSOLUTE MOTION

There is such another cause. That other cause is *absolute motion*, the motion of the Earth relative to the universe-wide absolute prime frame of reference, the effects of which have been neglected until now in the treatment of the behavior of the atomic orbital electrons. As was developed in Section 7, contrary to Einstein, there is an absolute frame of reference, an "at rest" frame. When the *Spherical-Center-of-Oscillation*'s oscillation is perfectly spherically symmetric then the center's velocity is zero and it is completely at rest. That is the universe's absolute frame to which all motion and all other frames are relative.

There exists throughout the universe a background radiation which is the residual radiation from the immense energy of the "big bang", the start of the universe. This radiation is, of course, relative to the absolute frame of reference. Measurements of Doppler frequency shift of this radiation due to the motion of the Earth relative to the absolute frame give an absolute velocity for the Earth of about *370 $^{km}/sec$*. The direction of the Earth's motion as indicated by those measurements is off in the direction from Earth of the constellation Leo.

The speed of the Earth in its orbit around the Sun is only about *31 $^{km}/sec$* so most of Earth's absolute speed is due to its motion relative to its galaxy, the Milky Way, and the absolute motion of that galaxy through space. Generally speaking it is likely that most if not all of the universe has a comparable magnitude of absolute velocity directed radially outward from the location of the original "big bang". But, whether or not, this *absolute* velocity of our Earth and our entire planetary-solar-galactic system of about *3.70·10^5 $^m/sec$ = 0.0012·c* must be taken into account in considering the behavior of the orbital electrons.

The most important factor in the stability of an atomic orbital electron is that it must not radiate energy. That requires that it experience no changes in the shape of its *Propagated Outward Flow* pattern of propagation forward, rearward and sideward. And, that requires that its speed remain constant. But, the speed of an orbital electron has two components: its orbital speed relative to the nucleus and its absolute linear speed because it is part of our overall solar system and galaxy.

In order for the electron to avoid radiating, it is its net speed, the resultant of those two components, which must remain constant. The way in which those two components combine to produce a net electron speed at any moment depends upon the orientation of the electron's orbital plane relative to the absolute velocity component of the electron, its atom and its solar-galactic system. The effect is illustrated in Figure 11-1, below.

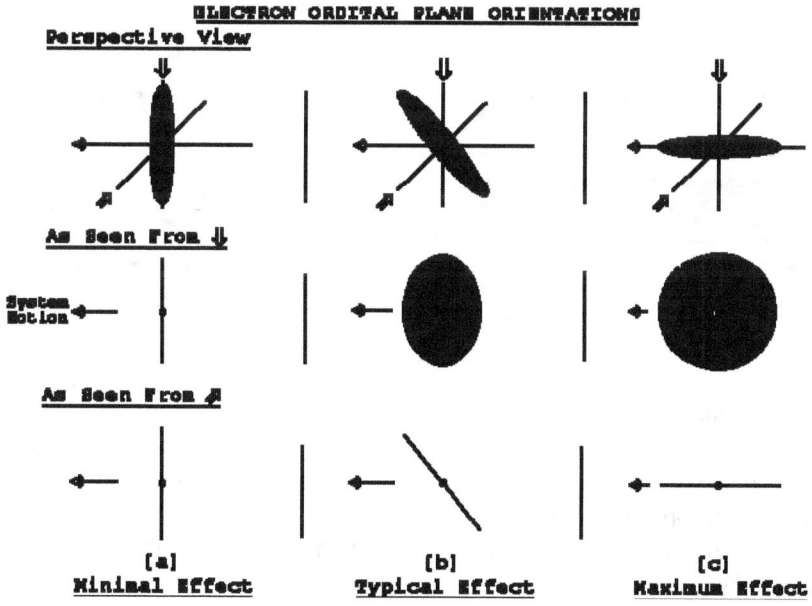

Figure 11-1
Relative Effect of Absolute Motion on Various Orbital Electrons

The figure illustrates different ways that the plane of an orbital electron's orbit can be oriented relative to the absolute motion of the overall atom. If the orbital plane is oriented at right angles to the direction of absolute motion, as in the *[a] Minimal Effect* column of the figure, then the absolute motion produces the same change in the overall electron resultant speed everywhere in the orbit. The electron's total speed is that resultant. Its orbital speed relative to the nucleus is the circular orbit speed for that orbital shell.

On the other hand, if the orbital plane is oriented parallel to the direction of absolute motion, as in the *[c] Maximum Effect* column of the figure, then the overall resultant speed of the electron varies between the sum of its circular orbital speed and the absolute motion speed and the difference of the two speeds (see Figure 11-2, below). In general, orbital planes are frequently oriented between those two extremes as illustrated in the *[b] Typical Effect* column Figure 11-1. For such cases the absolute motion can itself be resolved into two components: one at right angles to the particular orbital plan (Case *[a]*) and one parallel to it (Case *[c]*) and the resulting overall effect analyzed in terms of a combination of those two extreme cases.

Figure 11-2, below illustrates the analysis of the Case *[c] Maximum Effect* circumstances.

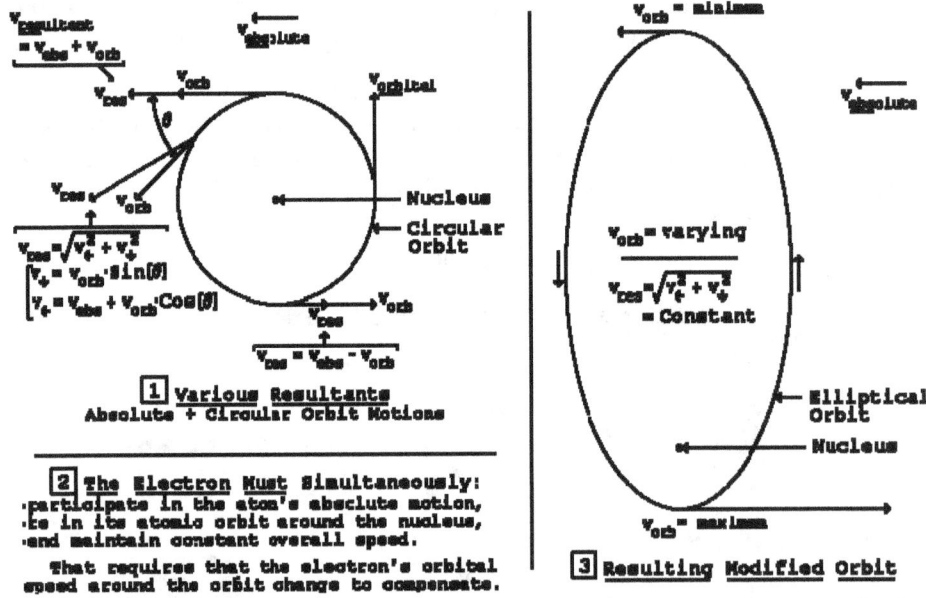

Figure 11-2

The figure is largely self-explanatory. If the electron is in a circular orbit (with consequent constant orbital speed) then the effect of the atom's absolute motion is to vary the electron's absolute speed, which is not acceptable. The only solution, the only *modus operandi*, is for the electron orbital speed to vary so as to compensate for the absolute motion and maintain constant absolute electron speed as shown in box 3 of the figure. The result is elliptical orbits for those orbits in which the orbital plane is not perpendicular to the direction of absolute system motion, that is for those orbits of Cases *[b]* or *[c]* of Figure 11-1.

The circular orbit speed in the $n = 1$ orbit of Hydrogen is about $2.2 \cdot 10^6$ m/sec. Our absolute speed is about $3.7 \cdot 10^5$ m/sec. The successive orbit speeds for $n = 2, 3,$ ... are $1/n$ times the $n = 1$ value. Thus the effect of absolute speed and the variations in orbital speeds are quite significant.

It is interesting to recall that the system of orbital quantum numbers developed by 20th Century physics and particularly elaborated by Dirac used the convention of the projection of an orbital angular momentum vector on a reference axis to define the various orbital tilts. It has now here been found that the "reference axis", an imaginary and missing element in traditional 20th Century physics terms, is actually the orbital plane orientation relative to the atom's absolute motion in space. The $l = 0$ value corresponds to the electron orbital plane being at right angles to the absolute motion, Case *[a]* of Figure 15-25. The $l = 1$ value produces a Case *[b]* situation. The horizontal orbit of Figure 11-2 is at right angles to the absolute motion and is circular. The other two orbits of the figure are now found to be elliptical, a pair tilted at equal but opposite angles relative to the absolute motion of the atom.

Returning to the problem of the cause of the *fine structure* in atomic spectra, there is a second consequence of the orbital electrons' absolute motion. Each electron has a component of magnetic field due to its straight line motion in space in addition to its orbital motion magnetic field. The electron's orbital magnetic field, which is perpendicular to the plane of the orbit, tends to align with the linear motion magnetic field that is due to the atom's absolute motion, which field is circumferential to the electron's direction of absolute motion. There are two possible alignment orientations, that is two orientations when there is no force acting that tends to change the orientation to one of the two. One is orbital motion in the same direction as the absolute motion magnetic field and the other is the opposite. The two differ slightly in energy. It is not "spin-orbit" coupling but "absolute motion - orbit" coupling that operates to produce the fine structure.

And that produces two "quantized" alternative values for the electron's angular momentum.

And that does away with any electron 'spin' and any electron quantized angular momentum in this kind of case.

High resolution spectral techniques, including the use of tunable lasers, disclose an even more closely spaced splitting of spectral lines which is called *hyperfine structure*. Analogous to the quantum mechanical explanation of fine structure in terms of hypothesized orbital electron spin, the hyperfine structure is attributed to hypothesized nuclear spin, its consequent magnetic field, and its interaction with the electrons. But, the nucleus can no more have spin and a spin magnetic field than can an orbital electron.

The hyperfine structure stems from electron orbital magnetic field interaction with the magnetic field due to the nucleus' absolute motion in space. Of course, overall the nuclear and orbital electron absolute motion magnetic fields cancel out since the direction of absolute motion is the same but the polarity of the moving charges are opposite. However locally, within the atom there is not general cancellation.

Recognizing the affect of absolute motion on the atomic fine and hyperfine line spectra eliminates quantized angular momentum and electron "spin" as factors involved in those phenomena.

Next, the phenomenon of the atomic orbital electrons' stable orbits.

Atomic Electrons Stable Versus Unstable Orbits

THE ATOMIC ELECTRONS STABLE ORBITS

The fact, observed at the time of Bohr's development of the relationship between atomic line spectra and atomic orbital structure, namely that the orbital lengths of the stable orbits of atomic electrons are an integer multiple of the orbiting electron's matter wave length was largely neglected because the problem of the matter wave frequency then remained unresolved. Furthermore, the Stern Gerlach experiment [see Section 13] had [erroneously] convinced the science community that quantized angular momentum was an inherent part of the nature of atomic particles.

The fact of the stable orbits has long been accepted without a specific reason, a specific operative cause, for those orbits and only those orbits being stable. It is not sufficient to assert that the stable orbits are those in which the orbiting electron' angular momentum is an integer multiple of $h/2\pi$ without supplying any cause or mechanism for that assertion. The matter wave of the orbiting electron now provides an operative reason, as follows.

For the orbit to be stable it must be the same for each pass, pass after pass. If each pass includes exactly an integer number of the orbital electron's matter wave lengths then each pass is the same in that regard. But if, for example, the orbital path length contains only $9/10$ of a matter wave length, $9/10$ of the matter wave period, then the next pass will contain the missing $1/10$ of the matter wave length or wave period plus $8/10$ of the next, and so on. The matter wave being sinusoidal in form, the successive orbital passes will be all different.

It is this behavior which operatively causes the "stable orbits", and only those orbits, to be stable. It has nothing to do with angular momentum nor quantization of angular momentum. For the angular momentum hypothesis there is no underlying reason nor mechanism to produce stability or instability. The quantization of angular momentum concept is merely an invented defined condition, without operative cause, just as were the "stable orbits" it seeks to explain until their being here justified in terms of the operative matter wave behavior

The statement that the orbital electron's angular momentum is quantized, as in the following traditional equation

$$(12\text{-}1) \quad m \cdot v \cdot R = n \cdot \frac{h}{2\pi} \qquad\qquad [n = 1, 2, ...]$$

is merely a mis-arrangement of

$$(12\text{-}2) \quad 2\pi \cdot R = n \cdot \frac{h}{m \cdot v} = n \cdot \lambda_{mw} \qquad\qquad [n = 1, 2, ...]$$

a statement that the orbital path length, $2\pi \cdot R$, must be an integral number of matter wavelengths, $n \cdot \lambda_{mw}$, long. The latter statement has a clear, simple, operational reason for its necessity. The former statement is arbitrary and is justified only because it produces the correct result, even if without an underlying rational reason.

It is this behavior which operatively causes the "stable orbits", and only those orbits, to be stable. It has nothing to do with angular momentum nor quantization of angular momentum.

How Electrons Are Forced Into Stable Orbits

With the vast amount of *Propagated Outward Flow* from myriad *Spherical-Centers-of-Oscillation* orbital electrons are continuously buffeted. How are specific stable orbit paths enforced? To analyze and quantify the deviations in the variable quantities involved, the radius, R, and the electron orbital velocity, v, will be expressed in terms of the orbit number, n, the number of matter wavelengths in the orbital path. That requires obtaining expressions for them that do not include any other variables.

That quantity, n, will here be deemed to be a continuous variable so that the R and v expressed in terms of n can be continuously variable and able to address locations between stable orbits, not merely the discrete amounts at the stable orbits.

The balance of forces for stability in a circular orbit requires

(12-3)   Centrifugal Force = Centripetal Force

$$\frac{m \cdot v^2}{R} = \frac{q^2}{4\pi \cdot \varepsilon \cdot R^2}$$

$$R = \frac{q^2}{4\pi \cdot \varepsilon \cdot m \cdot v^2}$$

(12-4)   Orbit Path Length = n · Matter Wavelength

$$2\pi \cdot R = n \cdot \frac{h}{m \cdot v}$$

$$2\pi \left[\frac{q^2}{4\pi \cdot \varepsilon \cdot m \cdot v^2} \right] = n \cdot \frac{h}{m \cdot v} \qquad \text{[Substitute } R \text{]}$$

$$v = \frac{q^2}{2\pi \cdot \varepsilon \cdot n \cdot h} \qquad \text{[Solve for } v \text{]}$$

$$v \propto \frac{1}{n}$$

(12-5) $$R = \frac{q^2}{4\pi \cdot \varepsilon \cdot m \cdot v^2} \propto \frac{q^2}{4\pi \cdot \varepsilon \cdot m \cdot \left[\dfrac{1}{n} \right]^2} \qquad \text{[Substitute } 12\text{-}4 \text{]}$$

$$R \propto n^2$$

In those terms the variation of the required centripetal force for a circular orbit as n varies is

(12-6)
$$F_{\text{Centripetal}} = \frac{m \cdot v^2}{R} \propto \frac{\left[\frac{1}{n} \right]^2}{n^2} = \frac{1}{n^4}$$

With constant charge the only variable in the expression for the Coulomb force is R in the denominator and is proportional to n^4. Therefore

(12-7) $\quad F_{Coulomb} \propto {}^1/n^4$

Thus the normal Coulomb force always provides the exact value of $F_{centripetal}$ required for a stable circular orbit.

The numerator of the Coulomb force expression is q^2. The variation from the force it exerts in the stable orbits depends on the ratio of the orbital path length, $2\pi \cdot R$, to the matter wavelength, $h/m \cdot v$. If that ratio is an integer then the behavior is the normal stable orbit Coulomb force.

If that ratio is not an integer then the force is *quasi-stable Coulomb*, as if the effective charge were modified as follows.

(12-8)

$$\text{Coulomb Force Numerator} \propto \frac{\text{Orbit Length}}{\lambda_{mw}}$$

$$\propto \frac{2\pi \cdot \mathbf{R}}{h/m\mathbf{v}} = \frac{2\pi \cdot \mathbf{R} \cdot m \cdot \mathbf{v}}{h}$$

$$\propto n^2 \cdot [{}^1/n] = n$$

$$\text{Coulomb Force Denominator} \propto \mathbf{R}^2 \propto n^4$$

and the overall *quasi-stable Coulomb* force then varies as

(12-9)

$$F_{Quasi-Coulomb} = \frac{\text{Numerator}}{\text{Denominator}} \quad \frac{n}{n^4} \propto = {}^1/n^3$$

The ratio of the quasi-Coulomb force to the normal Coulomb force then varies as

(12-10) $\quad \dfrac{F_{Quasi-Coulomb}}{F_{Normal\ Coulomb}} = \dfrac{{}^1/n^3}{{}^1/n^4} = n$

This means that for values of n somewhat larger than that of the next lower stable orbit integer value the actual Coulomb force acting, $F_{Quasi-Coulomb}$, is too large. For values of n somewhat below the stable orbit integer value the actual Coulomb force acting, $F_{Quasi-Coulomb}$, is too small.

Those results mean that:

- *Outside or above the stable orbit* integer value of orbit n the excessive values of $F_{Quasi-Coulomb}$ have the *net effect of moving the electron path inward*. The inward force produces an inward acceleration that is greater than the amount to produce a circular orbit. The excess acceleration produces inward electron velocity. (The inward $F_{Quasi-Coulomb}$ is greater than the outward "centrifugal force".)

- *Inside or below the stable orbit* integer value of orbit n the insufficient values of $F_{Quasi-Coulomb}$ have the *net effect of moving the electron path outward*. The inward force produces an inward acceleration that is less than the circular orbit amount. The deficiency produces less than circular motion, a net outward motion effect. (The inward $F_{Quasi-Coulomb}$ is less than the outward "centrifugal force".)

99

The overall effect is to force the electrons into stable orbits as Figure 12-1.

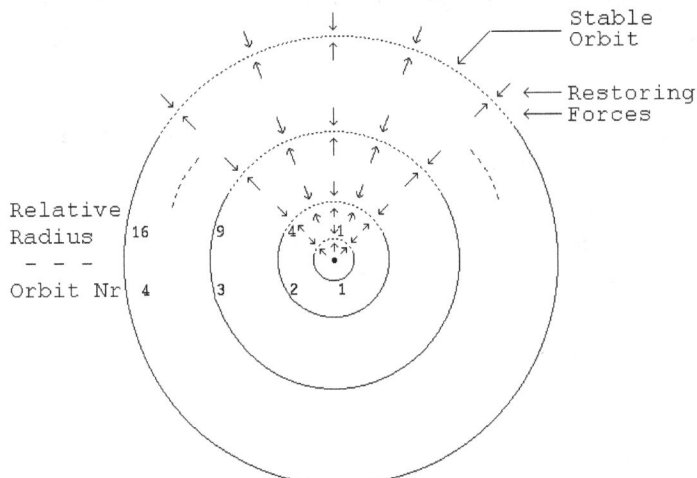

Figure 12-1
The Orbital Electrons Forced Into Integer Matter Wavelength Orbits

The Electrons' Transition Paths Between Stable Orbits

The above Figure 6-3 depicts the status when the orbital electrons are all in their lowest [least energy] orbits. When the outermost of those orbital locations is not occupied and the electron that should be in that position is in an excessively higher orbital location the action of the restoring forces is to direct that electron inward on an orbital transition path to fill the unoccupied position. That happens as follows.

The absence of an electron in the unoccupied position means that the positive electron-attracting field of the atom's positive nucleus is slightly un-offset by the orbital electrons' negative charges. With all of the lowest orbits filled the atom overall presents an electrically neutral status as viewed from outside, but with the outer electron missing that presentation is slightly of inner positive charge as viewed from the excessively higher orbital location electron.

That extra centrally directed attraction curves the pattern of restoring forces of Figure 12-1 to that of Figure 12-2, below.

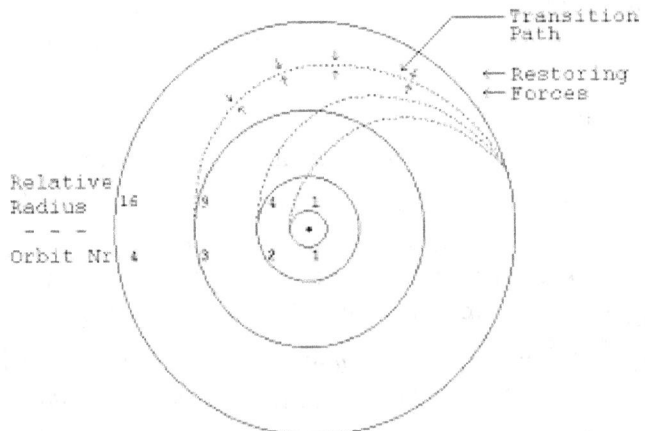

Figure 12-2
The Electrons Orbit Transition Paths

That drives the excessively higher orbital location electron inward to fill the empty location.

Any vacant location in the lowest energy positions of the orbital electron structure is automatically filled from above by this directing of the restoring forces. That is how an outer electron "knows" that there is a space that it can and should move into and that is how it follows the correct path to get there.

From any point in an outer orbit there is one specific path to each of the inner orbits of that outer orbit. Such paths, which involve inward motion of the electron in transition between stable orbits, have at each point in their path the correct inward motion to compensate for the deviation of the value there of $F_{Quasi-Coulomb}$ from what the normal Coulomb force should be at that point.

The electron velocity must vary smoothly from the stable velocity of the initial outer orbit through a period of increase and ending in the stable velocity of the final orbit. To do that without a discontinuity the variation must be in the form of a half cycle cosine. That is attested to by the sinusoidal nature of the *E-M* radiated photon. There can only be one such path that correctly compensates between any particular pair of initial and final orbits.

On either side of such a path the transition path restoring forces act just as for the stable orbits. The restoring forces arise because the stable orbit restoring forces will not allow locations between stable orbits.

Next, the most important, the Stern-Gerlach experiment

The End of Particle "Spin" and Its Quantized Angular Momentum

THE PROBLEM

The Stern-Gerlach Experiment

As the name suggests, particle spin was conceived as the rotation of a particle around an internal axis. This spin obeys the same mathematical laws as quantized angular momenta do. On the other hand, spin has some peculiar properties that distinguish it from orbital angular momenta.

Particles with spin can possess a magnetic dipole moment, just like a rotating electrically charged body in classical electrodynamics. Or, rather, individual particles exhibiting a magnetic dipole moment are deemed particles having spin. These magnetic moments can be experimentally observed in several ways, e.g. by the deflection of particles by inhomogeneous magnetic fields in a Stern-Gerlach experiment, or by measuring the magnetic fields generated by the particles themselves.

If the particle is treated as a classical magnetic dipole as it moves through a homogeneous magnetic field, the forces exerted on opposite ends of the dipole cancel each other out and the trajectory of the particle is unaffected.

However, if the magnetic field is inhomogeneous then the force on one end of the dipole will be slightly greater than the opposing force on the other end, so that there is a net force which deflects the particle's trajectory.

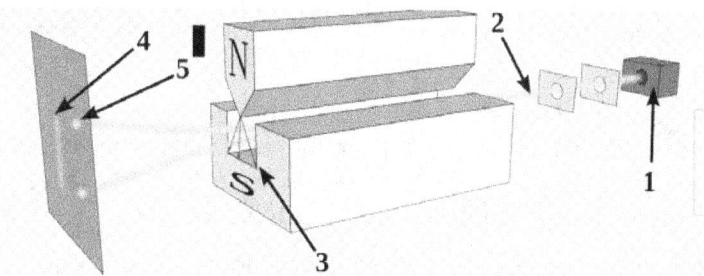

Figure 13-1
The Stern Gerlach Experiment

1 – Furnace   2 – Collimated beam   3 – inhomogeneous magnetic field
4 – Expected (not quantized) result   5 – "Quantized" result

103

The Stern-Gerlach experiment is normally conducted using silver atoms. That is because silver can be melted and then vaporized at a relatively low temperature. In the 103 experiment, Figure 13-1 above, silver is vaporized at a high temperature such that the atoms of the vapor are at high energy. Those atoms that escape from the furnace through a small aperture are collimated by further apertures into a narrow beam of silver atoms. The beam is then directed into an inhomogeneous magnetic field.

If the particles were classical spinning objects, one would expect the distribution of their spin angular momentum vectors to be random and continuous. Each particle would be deflected by an amount proportional to its magnetic moment, producing some density distribution on the detector screen. Instead, the particles passing through the Stern-Gerlach apparatus are deflected either up or down by a specific amount as shown in the figure.

This was taken to be a measurement of the quantum observable now known as spin angular momentum, which demonstrated possible outcomes of a measurement where the observable has a discrete set of values.

Historically, this experiment was decisive in establishing the physics concept of the reality of angular momentum quantization <u>in all atomic-scale systems</u>. It was the justification for the explanation of atomic fine and hyperfine spectra. It was the justification for contending that the stable atomic electron orbits were those for which the orbital angular momentum was an integer multiple of a fundamental angular momentum of $h/2\pi$.

The preceding Section 11 has demonstrated that quantized angular momentum and particle "spin" are not the correct explanation of the fine and hyperfine spectra and the just preceding Section 12 has demonstrated that it is integer multiples of the orbiting electron's matter wavelength, not angular momentum, that accounts for the stable electron orbits.

Now it shall be found that "spin" and quantized angular momentum are not what is involved in the apparently quantized result of the Stern-Gerlach experiment, as follows.

THE CAUSE OF THE STERN-GERLACH APPARENT QUANTIZATION

In the Stern-Gerlach experiment the metal silver is an excellent conductor of electricity. The reason is that its outermost orbital electron is very loosely bound to its atom. As a result that electron tends to become a free electron able to readily travel within the silver metal in which it is found. As the temperature and, therefore, energy increases in the quite hot and energetic silver vapor of the furnace the silver atoms of the collimated beam are mostly ionized, lacking that outer orbital electron.

Those positive silver ions flowing in a collimated beam constitute an electric current. And, in accordance with Ampere's Law that current results in a concentric magnetic field making each flowing silver ion a magnetic dipole. It is not "spin" nor a "natural property" of particles that produces the magnetic dipole, it is merely Ampere's Law and the ionized silver atoms.

But, the electrons lost by the ionized silver atoms are still present and flowing in the collimated beam. The beam is overall electrically neutral. That means that some of the silver atoms temporarily acquire an extra electron and become negative ions. And, the Ampere's Law magnetic field of the negative ion current flow produces a magnetic dipole directed opposite to that of the current of positive ions.

The collimated beam of silver atoms is a beam of magnetic dipoles of equal strength and opposite orientations. The strength of each is due to either one electron too few [for positive

104

ions] or one electron too many [for negative ions], a quantizing of the magnetic dipoles and therefore of their consequent deflection in the inhomogeneous magnetic field.

In brief the effect as if the silver atoms have "spin" and quantized angular momentum is due to their loosely bound outer electrons and their migration within the silver atom beam resulting in individual silver ions' local Ampere's Law current and consequent magnetic dipole.

THE END OF PARTICLE "SPIN" AND ITS QUANTIZED ANGULAR MOMENTUM

As presented earlier above, in discussions of Quantum Mechanics a property of particles identified as "spin" and involving angular momentum occurs frequently as for example a referring to "spin" up or "spin" down as quantum angular momentum 'states'. In those discussions it is often stated that no specific rotary motion (spin) is necessarily involved but that rather some intrinsic property of the particle being treated, an electron or an atom, is what is intended.

The intent is that quantized angular momentum is a natural property of particles such as electrons or atoms and that is contended in spite of there being no cause or mechanism for the particles to have that property and with the actual denial that any physical spin as rotation about a central axis is present.

That contention is defended by citing three different experimentally revealed behaviors:

- atomic spectra fine and hyperfine structure;
- atomic electron specific stable versus unstable orbits;
- the Stern-Gerlach experiment.

It has now been shown in Section 11 for atomic spectra fine and hyperfine structure, and in Section 12 for atomic electron specific stable versus unstable orbits, and in the current Section 13 just above for the Stern-Gerlach experiment, that there is no valid basis for the contending that quantized angular momentum is a "natural property" nor any valid basis for the general attribution of "spin".

THEN WHAT ABOUT "SPINTRONICS" ?

Spintronics is the study of the intrinsic spin of the electron and its associated magnetic moment. It has already been found here that fundamental particles, those which are Spherical-Centers-of-Oscillation, cannot spin and do not have "spin" as a "natural property".

However, the electron is a negatively charged particle that is always in motion. The most frequent appearance of its motion is in atomic electron orbits. But, free electrons are abundant, never at rest, always in curvilinear motion and that motion is effective as an electric current which results in a magnetic moment, which is the subject of spintronics.

Spintronics is not about a "natural occurring property" of electrons but merely the effect of it being a charge in constant motion acting per Ampere's Law. That is not a new Quantum effect, merely the classical action of classical physics. Any apparent quantization must be accounted for by some cause, some mechanism. It cannot appear without cause or mechanism as simply a "natural property".

105

CONCLUSION

The claims of Quantum Mechanics regarding Realism, Locality, and Entanglement as well as Quantized Angular Momentum have now all been addressed and resolved relying solely on the long known, long experimentally validated, classical physics which is characterized by being in full contact with material reality.

The primary lesson to be learned and applied to all of the pursuit of science is two-fold.

1 – In general valid scientific knowledge can only come about through investigating and understanding the causes of material effects, the mechanisms by which they operate.

2 – In the case of apparent effects for which cause and mechanism has not yet been developed it is valid science to pursue through all available means the investigation and search for validating cause and mechanism.

But, until that investigation and search has led to success the effects at issue must never be treated as if proven validated science which they cannot be.

Appendix A

Why No Immediate Mutual Annihilation

BACKGROUND OF THE PROBLEM

The Big Bang could only have resulted in equal amounts of matter and antimatter for the sake of the principle of conservation as presented in Section 1, *The Origin of Matter - Its Cause* with the assumption that there would have been a complete and almost instantaneous mutual annihilation.

Because that annihilation did not take place it has been hypothesized that the original symmetry was slightly skewed in favor of matter and that the universe is now all matter, all original antimatter having been annihilated with an equal amount of original matter. However that skewed balance conflicts with conservation in the Big Bang.

The Big Bang had to produce equal amounts of matter and antimatter and their total mutual annihilation did not occur because of the conditions there. Rather, while a moderate amount of initial matter / antimatter mutual annihilations may have taken place our present universe contains the remaining matter and antimatter in equal amounts, between some particles of which further mutual annihilations still occur at a modest rate.

The failure of comprehensive matter-antimatter immediate annihilation to occur develops as follows.

CONDITIONS AFFECTING MATTER / ANTIMATTER MUTUAL ANNIHILATION

What Is a Matter / Antimatter Annihilation ?

A positron-electron mutual annihilation, for example, is

$(A-1)$ $\quad _1e^0 + {}_{-1}e^0 \Rightarrow \approx + \approx$ where $\approx$ is a photon of gamma radiation

It happens as follows [per equation *2-6*].

$(A-2)$ $\quad (_1e^0) + (_{-1}e^0) = U_c \cdot [1 - \text{Cos}(2\pi \cdot f_e \cdot t)] - U_c \cdot [1 - \text{Cos}(2\pi \cdot f_e \cdot t)]$
$$= 0$$

The two oscillations literally cancel. The annihilation occurs because the two are point-by-point inverses of each other. Such an annihilation is depicted in Figure A-1 on the following page.

In general for a particular particle and some particular anti-particle of it, their phases and frequencies will not be identical because of their different velocities and histories of relativistic frequency shifts. However, for them to mutually annihilate they must remain co-located for some brief moment sufficient for the event to occur.

For the particles to be co-located for a brief moment their positions and velocities must be identical, which means that their frequencies and their phases will also be identical.

The mutual annihilation energy is the conversion into energy of the entire mass of the two particles involved. The mass of each of the particles is its oscillation [there is nothing else to be the mass]. At annihilation the two particles' oscillations cease to exist by cancelling each other out. Since the center oscillations cease, the last waves of *Propagated Outward Flow* are followed by no U-waves at all from those centers.

Section *5 – Ampere's Law* showed that E-M radiation is the propagation of changes in the *Propagated Outward Flow*, changes usually caused by velocity changes of charged particles. The ceasing at annihilation of the oscillations of the two particles involved [the largest change possible] causes a pair of gamma photons, equation *A-1*, to be propagated.

The photons carry off conservation maintaining energy and momentum. The frequency of each photon is the frequency of the oscillation that just ceased, which corresponds to the mass of the particle. In other words the photon energy, $W = h \cdot f$, is the energy equivalent of the entire mass of the of the particle annihilated.

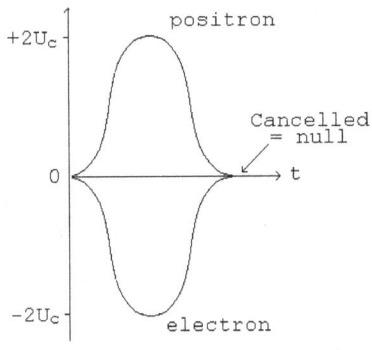

Figure A-1
A Mutual Annihilation

The first issue to investigate is the necessary conditions for a matter / antimatter annihilation to take place: how close must the particle and its antiparticle be and for how long must they remain in such sufficiently intimate contact ?

In addition to those two factors there is the more obvious requirement that the two particles involved be true antiparticles of each other [for example, a proton and an antiproton or an electron and a positron, but not a proton and a positron nor a proton and an electron]. Furthermore in general, particle / antiparticle annihilations are relatively unlikely between electrically neutral particles [for example, a neutron and an antineutron] because the only effects tending to bring the two together are their very weak gravitational attraction or chance encounter.

The Closeness Criterion

Indication of how close the two participating particles must be for their annihilation to take place can be found from the decay of a free neutron [not one that is part of an atomic nucleus] into a proton and an electron, a natural process with a mean lifetime before decay of about 881.5 seconds. For the neutron decay to be successful the proton and electron product particles must derive from the parent neutron not only their rest masses but also sufficient kinetic energy so that they are at escape velocity relative to each other, else they would be attracted back together and recombine. [One can neglect the also emitted electron anti-neutrino which is of zero or negligible mass.]

The escape velocity of the two particles is, at first consideration, an awkward problem because the separation distance of the two particles, which appears in the denominator of the expression for their Coulomb attraction, would seem to be required to be as small as zero. That is, at first consideration the escape velocity required is infinite. But, since infinite escape velocity is impossible yet the escape occurs, then the starting point, the minimum separation distance that can occur must be greater than zero. In other words, the neutron decay products, a proton and an electron, exist as such only when separated by some minimum Separation Distance, *s*, and their state at lesser separation distances appears as their parent neutron.

Therefore, since if the proton and the electron are separated by less than that minimum distance they do not exist as proton and electron but rather as the neutron, and at separation distances greater than that minimum they are the pair of separate particles, then that Separation Distance is a measure of how close a proton and an electron must be to unite into a neutron and is indicative of the spacing at which a particle and its antiparticle mutually annihilate.

The point is that the excess of the mass of the neutron over that of a proton plus that of an electron must supply the proton and electron relativistic kinetic masses needed to escape the decaying neutron. The detailed analysis and relativistic calculations can be found in Appendix A-1, *The Neutron*. The results are as follows.

(A-3) - The escape velocities:

$$v_e = 275,370,263. \quad \text{meters per second}$$
$$= 0.918,536,33 \cdot c$$
$$v_p = 379,350.6975 \quad \text{meters per second}$$
$$= 0.001,265,378 \cdot c$$

- The minimum Separation Distance:

$$S = 1.3 \cdot 10^{-15} \quad \text{meters}$$

Some years ago experiments involving measurement of the scattering of charged particles by atomic nuclei, yielded an empirical formula for the approximate value of the radius of an atomic nucleus to be

(A-4)   Radius = $[1.2 \cdot 10^{-15}] \cdot$ [Atomic Mass Number] meters

which formula would indicate that the radius of the proton as a Hydrogen nucleus (atomic mass number $A = 1$) is about $1.2 \cdot 10^{-15}$ meters.

The mass of the proton can be expressed as an equivalent energy, $W_p = m_p \cdot c^2$, and that as an equivalent frequency, $f_p = m_p \cdot c^2 / h$, or as an equivalent wavelength, $\lambda_p = c/f = h/m_p \cdot c$. That wavelength (not a "matter wavelength") for the proton is

(A-5) $\lambda_p = 1.321,410,0 \cdot 10^{-15}$ meters

111

quite near to the empirical value for the proton radius from equation *(A-4)* and the Separation Distance, S, of equation *(A-3)*. Thus the Separation Distance boundary between a proton and an electron as separate particles versus combined into a neutron is about *1* proton radius, the equivalent wavelength for the proton mass per equation *(A-3)*.

Then for a proton and an antiproton the boundary between their being the two separate particles and their mutually annihilating is a proton radius, a Separation Distance of $S_p = \lambda_p = 1.321,410,0 \cdot 10^{-15}$ *meters*. At that boundary if their velocities have a sufficient net component directly toward each other [per the time criterion, below] they would seem to be able, and likely, to mutually annihilate, and otherwise the annihilation would seem not possible.

Similarly, the mass of the electron or the positron can be expressed as the equivalent energy, $W_e = m_e \cdot c^2$, and that as its equivalent frequency, $f_e = m_e \cdot c^2/h$, or equivalent wavelength, $\lambda_e = c/f = h/m_e \cdot c$. That wavelength (not a "matter wavelength") for the electron / positron is

(A-6) $\quad \lambda_e = 2.426,310,6 \cdot 10^{-12}$ *meters.*

Then for an electron and a positron the boundary between their being the two separate particles and their mutually annihilating is a Separation Distance of $S_e = \lambda_e = 2.426,310,6 \cdot 10^{-12}$ *meters*. At that boundary if their velocities have a sufficient net component directly toward each other [per the time criterion, below] they would seem to be able, and likely, to mutually annihilate, and otherwise the annihilation would seem not possible.

Then, what is that sufficient net velocity ?

The Time Criterion

The mutual annihilation of a particle and its antiparticle is symbolized as in the following example for a proton and an antiproton.

(A-7) $\quad _1p^1 + {_{-1}}p^1 \Rightarrow \gamma + \gamma \quad$ where γ is a gamma photon

In the present case of a proton and an antiproton the mass of each of the protons is converted into the energy of the related γ photon. The frequency and period of each of those two photons is as follows.

(A-8) $\quad f_{\gamma p} = m_p \cdot c^2/h$

$\quad\quad\quad T_{\gamma p} = 1/f_{\gamma p} = h/[m_p \cdot c^2] = 4.407,749,3 \cdot 10^{-24}$ *seconds*

In communications theory it is shown that a sinusoidal oscillatory signal must be sampled at least twice per cycle for the signal to be correctly represented. That is, two independent datum's are required so as to determine the value of the oscillation's two absolute parameters, its amplitude and its frequency. [It's phase is relative, not absolute.] That implies that the time duration of a proton / antiproton mutual annihilation must be the period of each of the resulting photons.

(A-9) $\quad \Delta t_{proton\ /\ antiproton} = T_{\gamma p} = 4.407,749,3 \cdot 10^{-24}$ *seconds*

Similarly for an electron / positron mutual annihilation, the time duration would be

112

(A-10) $\Delta t_{electron\,/\,positron} = T_{\gamma e} = 8.093,301,0 \cdot 10^{-21}$ seconds.

While those are very brief times they are not instantaneous.

In the case of a particle and its antiparticle coming together from significantly far apart, the particles will have accumulated significant velocity toward each other by the time they arrive at Separation Distance s because of having been accelerated by their mutual Coulomb attraction. However, the situation was different for the Big Bang.

WHY THE CRITERIA FAILED IN THE CASE OF THE BIG BANG

The number of particles resulting from the original Big Bang is estimated to have been about 10^{85} [Appendix B, *The Limitation of the Original Envelopes*], and those particles emerged on paths that were initially radially outward. The event was overall spherically symmetrical on the large scale, but at the local particle level perfect symmetry was impossible because of the nature of finite particles versus a smooth non-particulate substance. Initially all of the particles were on divergent paths although for two adjacent particles the amount of the divergence was minute.

For a proton and an adjacent antiproton in the Big Bang to be separate [not annihilated] at the instant of being projected outward in the Big Bang, they had to be separated by at least the above-developed $s_p= 1.321,410,0 \cdot 10^{-15}$ meters. For them to then annihilate their Coulomb attraction would have had to accelerate them into co-locating in the required time criterion starting from their initially zero velocity toward each other. [Actually they would have had non-zero but minute velocities away from each other because each follows its own outward radial path.] The issue is whether their Coulomb attraction can accelerate the two particles to the point of co-locating within the time frame of equation $A-9$ [or equation $A-10$ for an electron / positron case].

If, for example, for their mutual annihilation, the proton or the antiproton is to travel <u>at constant velocity</u> its half of the separation distance, $\frac{1}{2} \cdot s_p$, in time $T_{\gamma p}$, so as to be co-located with its antiparticle at the end of that time, it would require a speed of

(A-11)
$$v_p = \frac{\frac{1}{2} \cdot s_p}{T_{\gamma p}} = 0.5 \cdot c \qquad \text{[half light speed]}$$

and if the electron or the positron, for their mutual annihilation, is to travel its half of the separation distance, s_e, in time $\frac{1}{2} \cdot T_{\gamma e}$ <u>at constant velocity</u> it would require a speed of

(A-12)
$$v_e = \frac{\frac{1}{2} \cdot s_e}{T_{\gamma e}} = 0.5 \cdot c \qquad \text{[half light speed]}.$$

The achieving of that speed, if even only by the very end of the extremely short time period of the acceleration and travel, 10^{-21} *seconds or less*, would be difficult. The particles moving continuously at that <u>constant velocity</u> throughout their travel from separated to co-located is impossible in that they commence their travel of distance s from essentially zero velocity toward each other.

Furthermore, the analysis of the Coulomb interaction at close separation distances presented in Appendix A-1, *The Neutron* shows that the attraction weakens drastically at close quarters per Figure A-2, below, reproduced from that appendix. [The figure shows the form of the reduction in the Coulomb attraction as a function of the charge separation radial distance relative to a proton mass equivalent wavelength, λ_p.]

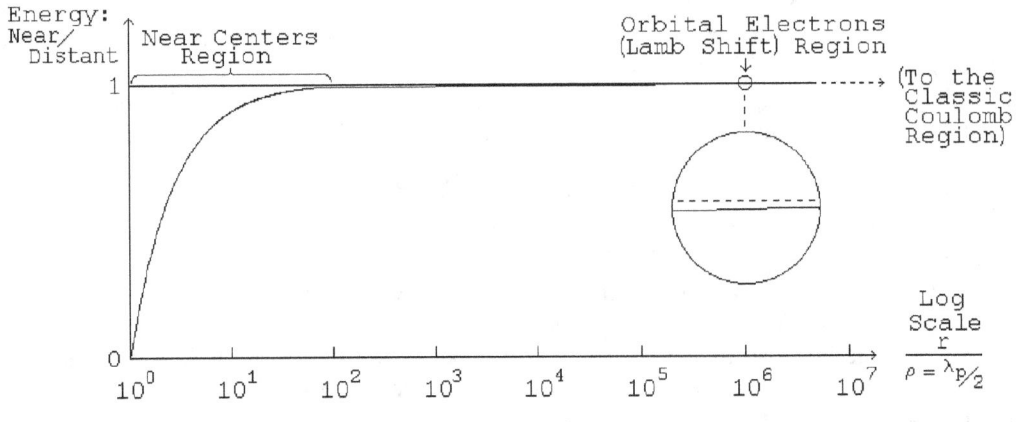

Figure A-2
Coulomb Effect <u>Reduction Factor</u> When Charges Are Near to Each Other

Finally, the posited particle and its antiparticle, emerging from the Big Bang, with spacing adjacent to each other as closely as possible, and on radially outward paths, were not alone. They were surrounded by a more or less uniform, symmetrical, large group of like particles and antiparticles. Any Coulomb tendency to unite the posited particle pair was largely offset by the similar tendency of each member to unite with the adjacent particle on its other side. The net Coulomb action on a specific particle or antiparticle was certainly insufficient to produce enough acceleration to enable the particle to transit its half of the Separation Distance in the required gamma photon period.

In summary:

- Adjacent Big Bang product particles and their antiparticles,

- Initially spaced optimally for co-locating [as closely as possible yet independently separate],

- Traveling outward at near light speed on essentially parallel paths [actually minutely diverging paths],

- Are unable to accelerate toward each other, from zero initial such velocity, quickly enough for their annihilation to produce the known actual gamma photons that would have to result from their mutual annihilation.

- That is, they cannot travel to the point of annihilation in time for the annihilation gamma photons to be the correct frequency to carry off the energy equivalent of the input particles, the pre-annihilation proton / antiproton or electron / positron.

In other words a Big Bang mutual annihilation was much more difficult, and rare, than one might have assumed. A large scale annihilation of matter and antimatter could not have taken place in the Big Bang. The result is that the present universe contains both matter and antimatter in equal amounts because of the original symmetry.

A Universe Containing Both Matter Regions and AntiMatter Regions

Why Matter and AntiMatter Regions Are Able to Co-Exist

Of course, matter / antimatter mutual annihilations in general are not as awkward as they were for the original Big Bang with its peculiar initial conditions. Of interest here, however, is the case of the interstellar medium. It is the interstellar medium that must be examined because it is the natural boundary between regions of matter and regions of antimatter; where, if they are to occur, the anticipated matter / antimatter annihilations should be occurring and yielding their looked-for gamma ray flux.

In the interstellar [and intergalactic] medium the particles and antiparticles start from being significantly separated, residing in the vacuum of interstellar space, which vacuum, while not devoid of competing particles, has a much lower particle density than the original Big Bang. They do not suffer the disadvantage of being in a dense milieu of particles and antiparticles whose Coulomb attractions tend to cancel out their effects. And, they avoid the disadvantage of always starting their mutual Coulomb attraction toward each other with no initial velocity. Without regard for any mutual attraction between particular particles and antiparticles, they all move with significant velocities.

However, those velocities are in general not oriented toward the combination of a pair. Rather, the velocity directions are a combination of [a] some component distributed randomly over the particles in essentially all possible directions, and [b] some amount corresponding to a general flow direction.

Table A-1, below summarizes the particle [and antiparticle where applicable] content of interstellar space. The density of the particles, and their related mean distance apart are such as to militate against any significant number of encounters, whether aided by Coulomb attraction or not. [Excepting solar wind, which is local to star's nearby environment, most of the interstellar medium is Hydrogen atoms, not ions.] [Gravitation can be ignored here, it being decades of orders of magnitude weaker than Coulomb attraction.]

Region	Size	Particle	
		Density [/cc]	Energy
Our Solar Wind	Sun Neighborhood	10.	0.001 - 0.004 $\times c$
Our Local Aloud	60 Light Years	0.1	~ 7,000 °K
Our Local Bubble	300 Light Years	0.001	~ 1,000,000 °K
Intergalactic Space	[The Universe]	0.000 ... ?	?

Table A-1 – The Interstellar Medium

As has been pointed out in analyses of our solar wind, with typically *1 atom* in each *10 cm³* of interstellar gas in our local cloud and *10 ions* in each *cm³* of our solar wind, the particles are so far apart that the solar wind and interstellar gas flow through each other without being disturbed by collisions. On that basis, the even less dense regions of the interstellar medium such as ones like our local bubble, those within galaxies in general, and those in intergalactic space are even less conducive to particle / antiparticle encounters.

Another factor bearing on the likelihood of matter / antimatter mutual annihilations occurring in interstellar space is as follows. Because gravitational and Coulomb field attraction communicate at c, particles are attracted to where the attractor was, not where it is. That tends to produce orbital motion or "sling shot" non-collision passages rather than direct collisions. For example, a proton traveling at $0.000,001 \cdot c$ *[only 300 meters/second]* and at a distance of 0.001 *millimeter* from another charged particle [compare that distance with the spacing implied by the densities of the above table] will travel a distance equal to *757 of its proton radii* during the time that its Coulomb field communicates at velocity c to the other charged particle its then Coulomb attraction impulse.

All of these various factors taken into account, matter / antimatter collisions must be quite infrequent events in the interstellar medium. When such mutual annihilations occur the appropriate gamma photons are emitted.

Indications of Some Matter / AntiMatter Mutual Annihilations

A most likely indication of our detection of cosmic matter / antimatter annihilations is Gamma Ray Bursts [GRB's].

GRB's are flashes of gamma rays coming from seemingly random places in deep space at random times. GRB's last from milliseconds to minutes, and are often followed by "afterglow" emission at longer wavelengths. Gamma-ray bursts are detected by orbiting [*Swift*] satellites about two to three times a week. All known GRB's come from outside our own galaxy. Most GRB's come from billions of light years away [as much as $z = 6.3$ or more].

Under the assumption that a given burst emits energy uniformly in all directions, some of the brightest bursts correspond to a total energy release of 10^{47} *joules*, nearly a solar mass converted into gamma-radiation in a small amount of time. No candidate process other than a significant matter-antimatter annihilation is able to liberate that much energy so quickly.

Appendix B

The Limitation of the Original Envelopes

This is to show how the otherwise infinite string of envelopes to the original oscillation at the start of the universe was subject to a finite limitation. By "finite limitation" is meant that in the vicinity of the cut-off number of envelopes, N_0, the amplitude of each of the further successive envelopes being imposed on the original $U(t)$ was successively significantly less than its immediate predecessor and the rate of that amplitude decrease increased sharply with further envelopes – there was a sharp cut-off of amplitude.

After a moderate number of such cut-off region envelopes the amplitude of any further envelopes becomes infinitesimal. While such infinitesimal (and still continuing to become ever more infinitesimal) envelopes theoretically go on to an infinite number of them, the result is equivalent to the convergence to a finite value of a mathematical infinite series such as, for example that of the cosine. The envelopes cut-off is a result of the mathematics of $U(t)$.

The key to that behavior is to be found in Table B-1, below, the expansion of the $Cos^n(x)$ function. The "Cosmic Egg" expression, equation $2-5$, repeated below

$(2-5)$ $\qquad U(t) = \pm U_0 \cdot \left[1 - Cos\left[2 \cdot \pi \cdot f_{env} \cdot t\right]\right]^{N_0} \cdot \left[1 - Cos\left[2 \cdot \pi \cdot f_{wve} \cdot t\right]\right]$

contains the factor

$(B-1)$
$$Cos^{N_0}\left[2\pi\left(f_{env}\right)t\right]$$

which creates the set of envelopes to the original oscillation. The expansion of the cosine raised to the power of its N_0 exponent behaves according to the pattern illustrated in Table B-1, below. Analysis of the patterns in the coefficients of the individual terms of the $Cos^n(x)$ expansion discloses a pattern related to the binomial expansion as demonstrated in the table.

119

(a) Binomial Expansion Coefficients $[a + b]^n$

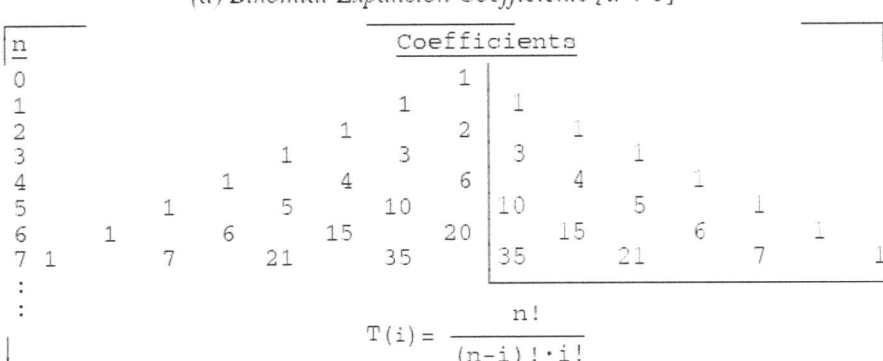

(b) $Cos^n(x)$ Expansion Coefficients

n	Times Cos(*), * =	0x	1x	2x	3x	4x	5x	6x	7x
0		1							
1		-	1						
2		1	-	1					
3		-	3	-	1				
4		3	-	4	-	1			
5		-	10	-	5	-	1		
6		10	-	15	-	6	-	1	
7		-	35	-	21	-	7	-	1
⋮									

$$T(i) = \frac{n!}{(n-i)! \cdot i!}$$

Table B-1

Clearly, with the exception of the constant term (where, in the table, * = 0x) the other terms of the expansion of $Cos^n(x)$ have the same coefficients as the corresponding terms of the binomial expansion. The formula for the binomial expansion can thus be used to obtain the coefficients for any value of n in the expansion of $Cos^n(x)$. in the present case for any value of N_0 in the expansion of the $U(t)$ factor

$$Cos^{N_0}\left[2\pi(f_{env})t\right]$$

The cut-off occurs around the value of N_0 regardless of what that value is. Therefore the value of N_0 is not important. Nevertheless it is of interest that various attempts to estimate it give values around 10^{85}.

$N_0 = 10^{85}$ is the n of the formula. It is not practicable and most likely not possible to calculate all of the coefficients of the cosine expansion of the envelopes for 10^{85} envelopes. On the other hand, it is not unreasonable to calculate the 85 cases corresponding to the frequency multiples of the expansion: 10^1, 10^2, 10^3, $\cdots$ 10^{85}.

Figure B-1, below, is a plot of the relative magnitude of the successive coefficients of the various frequency multiples $(1 \cdot x, \ 3 \cdot x, \ \cdots \ 10^{85} \cdot x)$, in the expansion of $Cos^n(x)$ for $n = N_0 = 10^{85}$. The plot indicates a sharp cut-off, an attenuation of the higher frequencies. Figure B-1(a) uses a linear horizontal axis and shows the cut-off in detail. Figure B-1(b) uses a logarithmic horizontal scale to better present the tremendous range in frequency multiples from 1 to 10^{85}. It shows that the cut-off is quite sharp and drastic.

This cut-off is merely the action of the mathematics of $\cos^n(x)$.

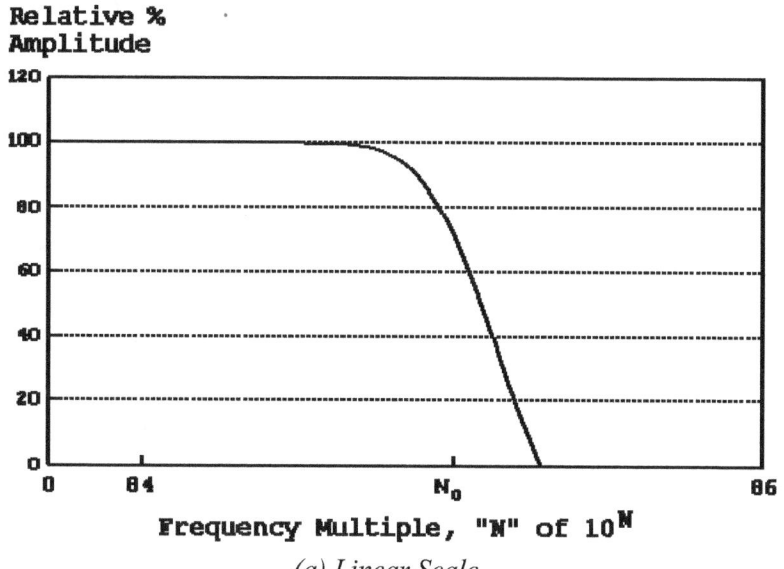

(a) Linear Scale

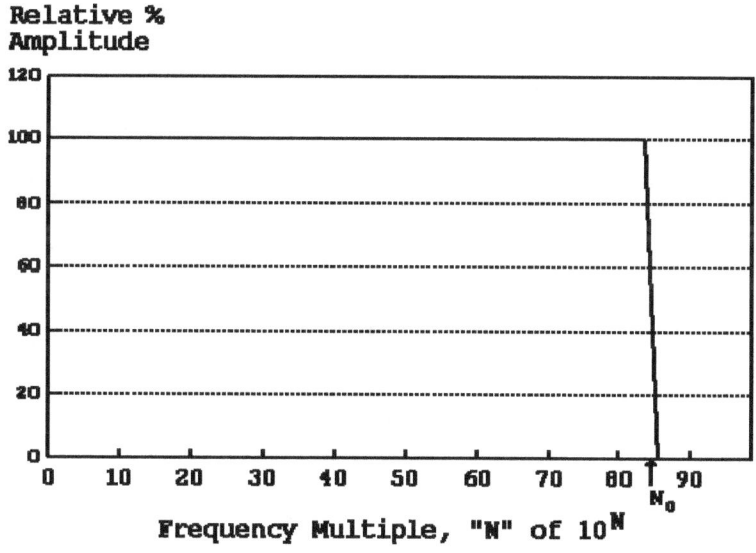

(b) Logarithmic Scale

Figure B-1
The Cos^n(x) Limitation of the "Cosmic Egg

121